BETWEEN HISTORY AND METHOD

W0107215

BOSTON STUDIES IN THE PHILOSOPHY OF SCIENCE

Editor

ROBERT S. COHEN, *Boston University*

Editorial Advisory Board

THOMAS F. GLICK, *Boston University*
ADOLF GRÜNBAUM, *University of Pittsburgh*
SAHOTRA SARKAR, *Boston University*
SYLVAN S. SCHWEBER, *Brandeis University*
JOHN J. STACHEL, *Boston University*
MARX W. WARTOFSKY, *Baruch College of the City University of New York*

VOLUME 145

STEFAN AMSTERDAMSKI

Institute of Philosophy and Sociology,
Polish Academy of Science

BETWEEN HISTORY AND METHOD

Disputes about the Rationality of Science

Translated by Olga Amsterdamska and Gene M. Moore

KLUWER ACADEMIC PUBLISHERS

DORDRECHT / BOSTON / LONDON

Library of Congress Cataloging-in-Publication Data

Amsterdamski, Stefan.
    [Między historią a metodą. English]
    Between history and method : disputes about the rationality of
science / Stefan Amsterdamski ; translated by Olga Amsterdamska and
Gene M. Moore.
        p.   cm.
    Includes bibliographical references and index.
    ISBN 978-94-010-5199-6 (alk. paper)
    1. Science--Philosophy. 2. Rationality.   I. Title.
    Q175.A5713   1992
    501--dc20   92-18172

ISBN 978-94-010-5199-6   ISBN 978-94-011-2706-6 (eBook)
DOI 10.1007/978-94-011-2706-6

Published by Kluwer Academic Publishers,
P.O. Box 17, 3300 AA Dordrecht, The Netherlands.

Kluwer Academic Publishers incorporates
the publishing programmes of
D. Reidel, Martinus Nijhoff, Dr W. Junk and MTP Press.

Sold and distributed in the U.S.A. and Canada
by Kluwer Academic Publishers,
101 Philip Drive, Norwell, MA 02061, U.S.A.

In all other countries, sold and distributed
by Kluwer Academic Publishers Group,
P.O. Box 322, 3300 AH Dordrecht, The Netherlands.

Printed on acid-free paper

Originally published by Miedzy Historia a Metoda, Państwowy Instytut,
Wydawniczy, Warszawa, 1983 under the title Miedzay Historia a Metoda.

All Rights Reserved
© 1992 Kluwer Academic Publishers and copyrightholders
as specified on appropriate pages within.
Softcover reprint of the hardcover 1st edition 1992
No part of the material protected by this copyright notice may be reproduced or
utilized in any form or by any means, electronic or mechanical,
including photocopying, recording or by any information storage and
retrieval system, without written permission from the copyright owner.

TABLE OF CONTENTS

PREFACE TO THE ENGLISH EDITION

In this book I have tried to develop further the ideas expressed in my previous work, *Between Experience and Metaphysics*, which was published in the same series in 1975.

Several years have passed since the original Polish edition (and then the Italian translation)[1] of this book appeared. The fact that the principal ideas expressed in it have withstood, as I see it, the brunt of criticism, has led me to remain basically with the original text. Two main changes have, however, been introduced.

First, I have added an Appendix containing the original version of a paper I presented at the Wissenschaftskolleg zu Berlin in June 1988 and a short postscript to that paper referring to comments made during two discussions at the Kolleg. Let me briefly explain the reason for this addition.

In recent years the landscape for historical and philosophical interpretation of the evolution of scientific knowledge has altered. The strongest of the new contenders for epistemological recognition are social constructivists, who analyze in detail how knowledge is produced within specific social settings, including the instruments and procedures of particular laboratories and the economic and political realities of particular scientific communities. The local character of these studies raises the question of whether they can ever provide generalizable epistemological claims. In fact, the proponents of the strong sociological program as well as the social constructivists believe that their (otherwise interesting) local case studies not only have epistemological consequences, but, what is more, that they compel us to change radically our opinions concerning the character of scientific knowledge and the mechanisms of its evolution.

So today we are no longer, as I wrote, in the same situation as T. S. Kuhn when he asked, "How could the history of science fail to be a source of phenomena to which theories of knowledge may legitimately be asked to apply?"[2] The tables have been turned, and after what I have written in my book against a purely methodological approach to the evolution of knowledge, I felt obliged to ask: Can social history and the sociology of knowledge indeed replace philosophy of science in solving epistemological problems? Can they, namely, explain the universalization of scientific knowledge, i.e. can they explain how claims to knowledge have come to be accepted beyond the local context in which the knowledge was produced and within a variety of settings where quite different practices, problems and political and social factors were at work? I am convinced that they cannot, at least not without accepting the idea of some theoretical, historically changing "background consensus" called in my book *the ideal of science*, within the framework of which scientific research is done, and which mediates the social and other circumstances of the "production" of knowledge.

The second change is less important. I have decided to omit Chapter VII of the original book, "Technical Rationality and the Ideal of Science." This was a polemic with certain concepts presented in the writings of the Polish philosopher Leszek Nowak, which had, as time has shown, a rather local character.

I am most grateful to professor R. S. Cohen for his interest in publishing this book in English. Besides the persons mentioned in the Preface to the Polish edition, I would like to express my gratitude to all those who have commented on the book in the course of discussions at the College de France (1984), the Wissenschaftskolleg in Berlin (1987/88), and at the faculty of Philosophy at Stanford University (California) in 1991—and especially to professors Nancy Cartwright (L.S.E.), Yehuda Elkana (Jerusalem), Peter Galison (Stanford), Horace Judson (Stanford). Timothy Lenoir (Stanford), René Thom (Bures sur Yvette), and Norton Wise (Los Angeles).

Finally, very special thanks to my daughter Olga and my son-in-law Gene Moore for their translation work; without their help this book would probably not have been published.

S. A.
Warsaw, January 1992.

PREFACE TO THE POLISH EDITION

This book is a continuation of the reflections presented in my previous book, *Between Experience and Metaphysics* (Dordrecht: Reidel, 1975). The present volume was written in 1975-1979, and now that its publication has become possible, I have made some minor additions and corrections.

In 1973/1974, thanks to a Fellowship from the American Council of Learned Societies, I was able to spend a year in the United States, where I had the opportunity to discuss my views with participants in the lively debates about the mechanisms of scientific development that were taking place at the time. Private conversations and public discussions contributed to my becoming aware that both sides in this debate shared a belief which was one of the sources of their disagreements and which, at the same time, made its resolution impossible. Both those who defended the idea that the development of science is a purely rational process which can be reconstructed historically on the basis of the rules of scientific methodology according to which it proceeds, and those who rejected this thesis, shared a conception of rationality which, in my opinion, was both historically and epistemologically dubious. Moreover, I became aware that in my earlier book I had to some extent also accepted this idea, and that as a result I had not been sufficiently consistent: I had to admit that many of the critical remarks about this issue were indeed correct.

Initially, the problem had appeared rather banal. I noticed, however, that when the criteria of rationality which find their expression in the scientific method are understood as a historical category, the debates about the development of science become largely irrelevant: for if we accept such a historical view of rationality, then it is possible to maintain *both* that the development of science is a rational process *and* that it is impossible to reconstruct this process historically on the basis of unchanging methodological rules expressing this rationality.

Such a purely semantic manner of solving this problem seemed suspicious, however: it was too simple. It suggested a manner of solving debates which I knew from elsewhere, and which was based on the possibility of changing the meaning of terms in such a way that two seemingly contradictory theses could be united by the conjunction "and" with no substantive consequences. And the reason why the debates continued despite such a simple possibility seemed obvious.

Pondering this issue, and reading yet again from this point of view both the texts which I already knew well and some newer work, I became convinced that this simple solution was after all not banal in the least, since its acceptance has a number of significant and often far-reaching consequences which cannot be incorporated into any of the previous competing positions. The fact that the conjunction allowed me to see a connection between the debates about the rationality of science and the debates about the role of science in modern culture appeared particularly significant to me. These debates are linked because the same thing is at stake in both of them: namely, the acceptance of a particular *ideal of science* which may be considered rational or irrational from the point of view of the realization of specific cultural values, and which also specifies the set of rules of investigation which are then, on the basis of a given ideal, considered rational.

As a result, many of the problems which concerned me until then began to form a coherent whole: methodological matters (such as the debates about the model of scientific explanation, the issue of the correspondence of theories, or the role of crucial experiments in the empirical sciences) as well as historical or sociological issues (such as the professionalization of science and its consequences, scientific revolutions, the role of scientism, etc.).

Many people and events contributed to this development: the discussions at the international seminar organized by the Aspen Institute in West Berlin, and a discussion meeting with T. S. Kuhn organized at the University of California at Berkeley in the Fall of 1976, as well as the colloquium which I conducted at the Institute of the History of Science, Education, and Technology of the Polish Academy of Sciences in the years 1975-1980. My collaboration with the *Enciclopedia Einaudi* (Turin), which forced me to specify in a relatively consistent manner my views on a variety of issues in the philosophy of science, also played an important role: the articles written for the encyclopædia are linked by the basic idea of this book.

I would like to express my gratitude to all those who on various occasions have helped me to write this book by offering critical remarks, and especially to J. Agassi. E. Chmielewska, B. Chwedeńczuk, R. S. Cohen. A. Grünbaum, G. Holton. T. S. Kuhn, W. Krajewski, J. Lalewicz. L. Laudan. E. Mokrzycki, E. Nagel, L. Nowak, B. Skarga, G. Stent, K. Szaniawski, M. Wartofski, and K. Wolicki. I must also mention two long and for me extremely interesting discussions with Imre Lakatos. I remember his interest in my ideas and the critical effort he exerted to make me abandon the erroneous direction in the philosophy of science which he was convinced I had chosen.

The most important debts to written texts are reflected in the references and the bibliography.

Stefan Amsterdamski
Warsaw, July 1981.

INTRODUCTION

The question of the rationality of science and its development, which is the subject of this book, is not one that scientists engaged in the pursuit of problems within their own disciplines are likely to encounter. Usually they have no doubts in this regard. They know the binding "rules of the game"; they know the conditions under which their colleagues will consider a proposed solution to a problem as a legitimate hypothesis, or possibly accept it as valid; and they know also that this acceptance will in turn confirm the rationality of the method they used to reach their solution. Rationality becomes problematic only when scientists begin to reflect about their own activities and those of scientists generally: about the goals and methods of these activities, and about their social functions and the roles played by their products in human life. In other words, the rationality of the mechanisms of scientific change and development becomes a problem in the philosophical—or more broadly, the humanistic—reflection on science, and must be considered in terms of the categories of such philosophical reflection, whether conducted by a scientist, a philosopher or an historian.

This humanistic reflection on science can proceed in two distinct, though related directions.

First, by examining science as an expression of human cognition, we might seek to discern in it our own nature, so as to gain a better understanding of ourselves as subjects capable of cognition, and of the various ways in which cognition can be shaped and conditioned. We can study science, as we do other products of human creativity, primarily in order to learn more about ourselves as its creators, and about the place of science within the totality of human culture. From this point of view, the issue of whether (and if so, the extent to which) such knowledge about science is necessary, or even useful in directing actual investigative processes or the real behavior of scientists, is of secondary importance. It has often been said that just as a knowledge of physiology never helped anyone to digest anything, so the organ of thought enabling us to interact with the world cannot stand to be applied to itself;[1] and so also the philosophy of science does not contribute to the making of discoveries. Even if this were true (and the statements of some scientists seem to deny it, as do the claims of philosophers, who in this manner are trying, typically enough, to advertise

1

their wares), the direction of such reflection would not thereby be rendered invalid. Nowhere is it written that the only justification for such reflection is its usefulness for scientific inquiry. It can fulfill intellectual needs other than merely serving the subject of its attention.

Secondly, the reflection on science, especially when it assumes the form of a methodology, may—and in fact often does—aim to formulate rules of investigative behavior and evaluative criteria for judging its results. Such rules and criteria are then said to lead to cognitive success. Although this reflection assumes a clearly pragmatic character, it must nevertheless rely on some, if only implicit, view of science and its goals, for otherwise it would be unable to supply any rules of inquiry. *But the moment this conception of science is put in doubt for any reason, the methodological rules derived from it will also be rendered problematic.*

Despite their different orientations, these two types of humanistic reflection on science, both conducted in our culture since times immemorial—the first of which I would call *philosophy of science* and the second *methodology of science*—are not really independent of each other, and they feed on each other's results. Nevertheless, as we shall see, the problem of the rationality of science does not arise in the same manner from both perspectives. From the first perspective, the problem of rationality concerns the functions of science in culture, and the evaluation of these functions in terms of broader cultural values; while from the second perspective, the rationality of science is treated most often as unproblematic and indisputable, and the problem is defined rather as one of articulating effective methods for the implementation of these values.

One of the basic aims of this volume is to show that these two clearly distinct conceptions of rationality cannot be separated from one another in the reflection on science.

For many years science has been treated as the embodiment of human rationality. It was seen as a feature specific to our culture, and its development was represented as the result of a systematic application of the rational method of investigation. It is of course true that philosophers have long argued among themselves about what this method consists of, and what it should consist of (in philosophical disputes, the descriptive and the normative aspects always appear together); but no one questioned the rationality of science and its development. Even the romantics who were opposed to rationalism as an intellectual attitude towards the world saw science as the major bastion and foundation of this attitude; and when they criticized science it was precisely because of its "cold rationality," because it "stripped all value and quality from the world," and replaced them with numbers and measures. "Feeling and faith speak stronger to me than the sage's glasses and eyes," as the Polish romantic poet, Adam Mickiewicz, declared.

Despite epistemological divergences on the issue of how rational cognition is possible, and what its method should be, and despite axiological conflicts about the role of science in our culture (conflicts to which we shall return, but which scientists themselves did not consider important until the end of the nineteenth century), a general consensus has continued to provide a framework for these disputes. This consensus was marked by a conception of human rationality and an ideal of scientific investigation which made the question of the rationality of science impossible to pose. Posing such a question would have been about as sensible as inquiring whether it is true that every bachelor is an unmarried man. Scientificity and rationality vouched for each other.

Today this situation has changed, and the question of the rationality of science and its development has become one of the most controversial topics in the philosophical reflection on science.

The causes of this change—which are discussed in chapters IV and V—are, I believe, twofold. Today, for the first time in three hundred years, we are no longer morally certain that the development of scientific knowledge and the technological progress linked with it are indeed unequivocally beneficial. Moreover, unlike the scientists and philosophers of a century ago, we no longer possess the conviction that scientific knowledge can be fully objective, that it can be an unmediated product of an autonomous knowing subject, and that its history is simply the history of Reason. The rationality of science is thus called into question both because of the effects of scientific development on our culture, and because of the processes of the production of scientific knowledge leading to these effects. As we shall see, the development of knowledge has undermined the idea of the cognitive autonomy of the subject, which was the essential precondition for the conception of the subject's rationality, at least in science.

Some critics of science claim, and not without reason, that over the last three hundred years, and especially since the beginning of this century, science has become increasingly involved with technologies for manipulating persons and things, and with a technocratic social order. Moreover, they claim that this connection is not accidental, but rather a necessary consequence of the scientific method. According to Herbert Marcuse, for example, "science, *by virtue of its own method* and concepts, has projected and promoted a universe in which the domination of nature has remained linked to the domination of man—a link which tends to be fatal to this universe as a whole."[2]   "[A] dictatorship of test tubes rather than of hobnailed boots will not make it any less a dictatorship."[3]   According to Habermas, science is the ideology of a technology which has become alienated from the power of human reason.   And Paul Feyerabend asks, "What's so great about science?"[4]

The critique of technocratic systems ("systems" since technocracy can exist under different political regimes) turns into a critique of science as an activity and as a social institution supporting such systems, or even calling them into being.   And since these systems can be considered irrational in terms of the realization of specific cultural values, science is also seen to be serving such irrational goals—or, to be precise, goals which are not always strictly rational.   The rationality of science is evaluated here with respect to its cultural functions and in view of its role in creating and destroying values in social life.

This type of criticism is clearly a continuation of the nineteenth-century romantic trend in European culture to which we alluded earlier. It is also called neo-romantic.   It criticizes our science-based industrial society for destroying the natural community of mankind, dissolving traditional human ties and reducing man to the condition of a thing subjected to anonymous, hostile, and autonomous powers.   Motivating this critique is the dream of eliminating all mediation between the individual and the community, or between the individual and nature, a fantasy of the return to a state of original grace.

If, however, scientists a century ago were justified in remaining indifferent to this type of critique, or dismissing it as a longing for a lost paradise which never existed in fact, or as an apotheosis of the "noble savage" unspoiled by civilization, today such a dismissal would be disingenuous.   The twentieth century has deprived scientists of their moral certitude that by engaging in scientific research they can only benefit mankind; and by the same token, it has undermined the assumption that merely following methodological rules constitutes a sufficient ethical code for a scientist.   As Robert Oppenheimer said after the Congressional hearings, "Physicists have now learned about sin."   As a result, scientists are now called upon to face moral conflicts from which, at least subjectively, they were free until quite recently.   They have ceased to enjoy a privileged position in the world.   Bitter skepticism concerning the objective validity and the moral value of scientific knowledge, and of the changes brought about by its development, is an important part of today's intellectual climate.   Various social utopias can still—more or less effectively—try to render human consciousness insensitive to these conflicts, but they can no longer eliminate or ignore them.

In this situation the focus of humanistic reflection on science has also had to change.

> While scientists a hundred years ago gained self-knowledge simply by methodological reflection, and could participate in humanist culture by communicating the results of these reflections to others, that is, by transmitting information about scientific theories and discoveries within the context of the problematics of philosophy—today, they achieve self-knowledge not so much by reflecting about the methodology required for developing new theories and making discoveries as by reflecting about science as a social phenomenon, and they can participate in humanistic culture by attempting to understand the role of science in contemporary life and the ambiguous perspectives opening up through its continuous development.[5]

Precisely this change has led to a great interest in the history and sociology of science and to the confrontation of their results with the theory of knowledge. This interest has led to a questioning of the thesis of the rationality of scientific development in other respects as well.

Many investigators attempting to examine the mechanisms of the development of scientific knowledge (often called the logic of scientific development) are questioning the received view that the development of science takes place, or can take place, as a result of the exclusive application of investigative procedures regarded as rational and codified in various ways by scientific methodology. In other words, what is being criticized is not just a few specific methodological conceptions but a more general thesis, namely the notion that any methodology can, even if only in satisfactory approximation, represent the mechanism of the actual historical process, because every such methodology by definition excludes the extra-methodological factors which condition this process, treating them as "external," "non-essential," "accidental," or "non-rational." By reconstructing the development of science on the basis of accepted methodological models (the logic of development), these methodologies falsify history and create a false view of the rationality of this process.

This view—formulated and justified in a variety of ways in many works, especially those of Thomas S. Kuhn, Paul K. Feyerabend, Yehuda Elkana, Stephen Toulmin, and Michael Polanyi—has far-reaching historiographical and methodological consequences.

If, on the basis of historical analyses, one denies the existence of any ahistorical methodological rules whose consistent application would lead to the actual development of science (in the common-sense meaning of the term), one thereby undermines the thesis of the rationality of this process in several essential points.

First, one is then implying a basic discontinuity in history, that is, the occurrence of transformations (revolutions) which cannot be described in purely methodological terms. This is the case because the previously accepted and applied methodological rules are themselves among the things that change and require explanation in such transformations. In order to explain such transformations one has to appeal to external—sociological, historical, and psychological—factors. When the criteria of rationality are identified with certain universally valid rules whose only justification is epistemological, and which the methodology of science is supposed to formulate, then it turns out, by the very same token, that this process itself is not fully rational.

Secondly, this leads to the conclusion that the discovery of the mechanisms of the development of knowledge requires us to go beyond a so-called rational reconstruction (which is based precisely on the representation of this process as unfolding according to consistent and rational rules for the acceptance or rejection of claims, the construction of theories, or the explanation of phenomena). It then becomes necessary to take into account the influence of psychological, historical or sociological factors which—in terms of the accepted concept of rationality—appear to be irrational or non-rational. The history of science, it is said, is not the realization of the evolution of autonomous Reason. The problem of understanding various historical periods in the development of scientific knowledge is basically—despite the claims of ahistorical methodologies—analogous to the anthropological problem of understanding other cultures, and involves analogous theoretical difficulties.[6] Thus, every "rational" reconstruction misrepresents the process of development and creates a false image of the place and status of science in culture.

Thirdly, since the development of science does not result from the application of a consistent rational investigative method; and since at moments of discontinuity further development depends in part on the acceptance of new "research programmes" (Lakatos) or "paradigms" (Kuhn) which do not follow logically from previously accepted programmes or paradigms, but which lead to at least a partial incommensurability between new and old theories, and, most importantly, impose their own rules of investigation; then it can be said that a given methodology is unable to formulate any conclusive criteria of choice between competing theories. Under these conditions, the regulation of cognitive activity through the formulation of methodological criteria which are expected to indicate the rational course of action (the problem of philosophy of science ever since the times of Bacon and Descartes) must indeed play a conservative role with respect to the current state of knowledge. Methodology would then petrify the existing ideal of science and the investigative methods it contains, presenting them as the only rational or even the only ethical ones.[7] This is the point at which the critique of the thesis of the rationality of science as guaranteed by its method converges with the critique of science because of the effects which follow from the application of this method. Philosophy, which defends the thesis of the rationality of science in this sense, does not allow for a change in its ideal—which is irrational from the point of view of cultural values; and by the same token it plays an ideological role, defending the status quo, the existing social order served by science conducted in this manner.[8]

If the claim with which I began is correct, that is, if indeed the question of the rationality of science and its development could not be formulated on the basis of the modern conception of rationality and the ideal of science formed in European culture in the sixteenth and seventeenth centuries, both of which have defined the goals and directions of scientific effort for roughly three hundred years, and if today this question is the subject of fundamental controversies, then we might be justified in suspecting that this conception of rationality or the ideal of science itself has become problematic. Only this could explain the fact that an assertion whose truth has not been questioned for decades, simply because of the meanings accorded the terms "scientific" and "rational," could become a subject of controversy which cannot now be resolved on the basis of a semantic analysis of the terms in which it is stated. Since in the past the rationality of science and its development seems to have been warranted in an *a priori* manner, while today it appears as a controversial problem, then one has to examine in detail the nature of the concept of rationality and the ideal of science that are now being called into question.

I believe—and the attempt to demonstrate this is the second major aim of this book—that the concept of rationality is not purely epistemological and descriptive, but that it is historically conditioned and evaluative, and that *the models of rational investigative behavior reconstructed by the various methodologies are imposed by the accepted ideals of scientific knowledge.* Accordingly, the first chapter of this book tries to explain what is understood by the notion of an *ideal of science* and to reflect on its functions in the development of knowledge. The second chapter attempts to demonstrate how methodological disagreements are conditioned by the acceptance of distinct ideals of scientific knowledge, on the basis of an analysis of the controversies surrounding the rules of scientific explanation (any other methodological problem could also, of course, have served as an example). In the third chapter I attempt to reconstruct the modern ideal of science in order to discuss the sources and causes of its present crisis. I think that only on this basis can we come to a better understanding of the contemporary controversies surrounding the rationality of scientific development, to which the final sections of this book are devoted and which have been mentioned briefly above.

CHAPTER I

THE DEVELOPMENT OF KNOWLEDGE
AND THE IDEALS OF SCIENCE

1.

Every human society possesses an extra-genetic means of transmitting knowledge from generation to generation. Accordingly, every society, or at least every society characterized by division of labor, must include groups of people and institutions whose task it is to cultivate—to gather and transmit—knowledge. Of course, this circumstance does not in itself determine either the character of these groups and institutions or the types of knowledge which they cultivate; but if the knowledge provided by such a group were not valued for one reason or another by at least a part of the society, the group or institution would be unable to maintain a status allowing it to exist and act.[1]

Various groups of men of knowledge—priests, magi, sages, experts in various areas of practice—have been able to cultivate different kinds of knowledge. For example, they could focus primarily on preserving the knowledge inherited from previous generations, protecting it from all sorts of novelties threatening to alter patterns of social life which have become sanctified by tradition. Or such groups could strive systematically to expand and enrich knowledge. They could see themselves as repositories of a secret wisdom, whose possession requires some kind of initiation or some special qualities on the part of those who can achieve it. Alternatively, such groups could feel that the knowledge they provide should be accessible to everyone, and that in this sense it is universally valid. They could search first of all for answers to cosmological questions concerning the universe and the place of man in it, or see their task primarily as that of supplying prescriptions for effective action. They could also try to achieve both of these goals at the same time, assuming that a knowledge of the cosmic order, and insight into what is necessary, possible, or impossible within this order, is a necessary prerequisite for effective action. Such groups could also legitimate the validity of the knowledge they provide, and ascribe different values to different methods for its acquisition, justification and testing. Reflecting on their own activities, such groups could inquire into the relations among various types of knowledge and value them according to their own particular criteria; in other words,

9

they could engage in philosophical reflection.[2] The internal organizations of these groups and their relations with the rest of society have always depended on the types of knowledge which they sought to attain, and on the overall structures of the societies in which they functioned.

There is no doubt that with the growing division of labor and differentiation of social roles, these groups of men of knowledge also became—at least in some societies—more differentiated and autonomous. They began to form their own ideas of valuable knowledge while rationalizing their own activities and legitimating the social status they either enjoyed or wished to enjoy. In other words, these groups themselves began to develop their own group ideologies. Thus, the development of various branches of knowledge—magic, religion, cosmology, technical knowledge, etc.—would result both from the need to gather and transmit knowledge that is common to all societies, and from the acceptance of particular group conceptions of what constitutes valuable knowledge. Thus, a historian or sociologist might well ask why a certain type of knowledge was considered valuable in a particular society, instead of attempting to evaluate this knowledge on the basis of the models of knowledge accepted as valuable in the society or group to which the historian or sociologist happens to belong. The changes that take place in these models of evaluation, their filiations and mutations, are elements of the history of knowledge which are just as important as the history of any of these models separately.

What we today consider to be science is undoubtedly, at least genetically, a kind of derivative of the types of knowledge inherited from the past, some elements of which have by now become less important or even disappeared, while others are still being pursued. Contemporary scientific communities have their distant ancestors in various groups of men of knowledge. However, when we pose the questions of when, where, and how science was "born," and how it differs from other types of knowledge, there seems to be no way in which such questions can be answered unequivocally and legitimately. And this is so not because we lack sufficient historical information; the root of the difficulty lies elsewhere. The difficulty is that the answer to this question always depends on the ideal of knowledge which we accept when we answer it.

Those who treat science as a disinterested search for universally valid truths will see its origins above all in the cosmological and philosophical investigations of the ancient Greeks. Those who see it primarily as the ability to manipulate objects effectively, or the art of controlling the natural and artificial environments of man, can trace its beginnings to the technological abilities of various tribes, or may note its similarities with magic, or even claim that science has always existed, as long as human society itself, since every society must possess some—even if only rudimentary—technology. Those finally who claim that the essence of scientific knowledge is contained in its exact mathematical character, or that there is only as much truth in science as there is mathematics, might stress the special role of the mathematical achievements of the Babylonians. And these are probably not the only possibilities, since the view that science is a creation of modern European culture born in the sixteenth and seventeenth centuries is also quite widespread.

Providing answers to the questions posed above is, in other words, comparable to deciding which of the many tributaries of a river constitutes its true source. Why should we not trace the source of the river which flows into the Baltic Sea at Gdańsk to the Bieszczady Mountains rather than to near Barania Góra? Is it because if that were so, Cracow would not be a city on the Vistula, but merely on one of its tributaries? If the location of a city on the Vistula River were associated in the minds of its inhabitants with some particular value, then it is likely that the inhabitants of Cracow would indeed object to this terminological alteration, while the inhabitants of Przemyśl would perhaps find it rather to their liking.

A historian—regardless of whether he is a historian of science, art, or literature—is, in this sense, always travelling upstream from the "estuary" where he finds himself to the "sources" which are being investigated. He is looking for the past in terms of how he sees and evaluates the present. Thus, whether we locate the beginnings of science in the mathematical achievements of the Babylonians or in the technological abilities of the Chinese, in the philosophical and cosmological speculations of the Greeks or only in the mathematical and experimental methods of Galileo—we are always choosing a tradition. *And we make our choice on the basis of a conception of valuable knowledge which for one reason or another we accept, and which we call science.*

Just as an accepted ideal of science shapes the answer to the question of where and how science was born, so also this ideal decides what types of knowledge are to be classified as scientific. There can be no doubt that in our culture this classification has a normative character.

Historians, philosophers, or sociologists of science do give answers to questions posed in this manner. They either do so explicitly, by formulating various definitions of science, citing some criteria of demarcation which will distinguish science from other types of knowledge or other types of cognitive activity, or else they accept such answers implicitly. This is fully understandable, because otherwise they would be unable to investigate those aspects of science which interest them. Nevertheless, we have to be aware of the fact that every decision of this sort is always, whether explicitly or implicitly, relative to the accepted ideal of scientific knowledge, and it constitutes a projection onto the past, or—possibly—onto the future. Thus, every such decision, if it pretends to be universally valid, is in effect the selection of a tradition corresponding to a given idea of science from the total history of human cognition, and the legitimation of a tradition which assumes that a particular ideal of science is something obvious, unproblematic, and the only one possible.[3] There is no other way to provide an answer to this question. *The history of science, its traditions, sources, and potential boundaries, are always constituted by some particular ideal of scientific knowledge accepted at a given historical time by a specific group of people.* Some of these ideals might be socially accepted and institutionalized at a given time, and direct the cognitive activities of scientists; while others might exist as individual or group ideas which enjoy no such social approval and, at least for a while, might not be considered historically productive.

Thus, despite the various attempts of methodologists searching for universally valid criteria of demarcation, science can be distinguished from other types of knowledge in an historically adequate manner only conventionally and normatively: conventionally, because it is a matter of convention to regard a given ideal as universally valid; and normatively, because the choice of any convention is normatively conditioned. And this is so independently of whether the term "science" is used to refer to a certain kind of knowledge, a certain kind of cognitive activity by which such knowledge is produced, or the social institution in which such activity is conducted. Both the definitions which are accepted for regulative purposes and those accepted for the purpose of describing science are necessarily conventional and normative. When such definitions pretend to universal validity they simply conceal their conventional character. By accepting these definitions we construct the realm of those phenomena which can be considered scientific, and we define the historical and intellectual boundaries of the phenomena. It is obvious that not everything that has ever been or might in the future be considered as science can be accommodated in such conventional definitions.

This is why I believe that all philosophy of science and all methodology has an unavoidably normative character, and that the drawing of a distinction between descriptive and normative methodology is (in this respect) a misunderstanding which obscures the fact that a silent acceptance of some normative idea of science is a necessary condition for the formulation of all descriptive methodology.

From the perspective presented here, it is difficult to foresee the possibility of any objective solution to the question of whether science is a creation of modern European culture, or whether various historically changing ideals of science have succeeded one another within our cultural tradition. This is so because we again encounter exactly the same problem as before, when we attempted to define the boundaries of "our" culture. In the words of Leszek Kolakowski:

> The changes which constantly take place in this storehouse [of culturally accepted values] do not preclude the existence of a certain deposit which can be preserved for an extremely long time next to the moribund elements, and which allows us—not without doubts, of course—to establish a multi-generational and multinational cultural continuity stretching for centuries. Some such thesaurus has always legitimized the historian's claim that the continuity between the Ancient Mediterranean and Christian culture, and between these and the culture of the contemporary industrial societies which emerged from the Latin Middle Ages, is so extensive that we are actually dealing with one diachronic society. This continuity joins together a set of values which, despite gradual and mutational changes of all sorts, have prevailed from the times of ancient Greece.[4]

The doubts hinted at in this statement regarding the means of establishing this continuity are surely connected with the fact that the storehouse of cultural values may change not only in terms of the reorganization of a given inventory, but also because there is not a single value the demise of which could be considered equivalent to the death of the whole culture, nor could there be a change so extensive that we could no longer claim that we are still dealing with the same culture. In other words, we can posit nothing beyond genidentity.

Precisely this circumstance shapes all our ideas about the "end" or the "crisis" of our culture, as well—and with equal justification—as all our ideas about its immortality. All disputes concerning such issues can be resolved only normatively and conventionally: conventionally, because how we decide that the death of a given value constitutes the death of a culture is a matter of convention; and normatively, because the choice of such a value can only be sustained normatively. Such is the inevitable character of all demarcations of "our" culture, because "we cannot hold our history in our minds without any landmarks, or as an ocean without fixed points, and we may talk about this civilisation and that as though they were ultimate units, provided we are not superstitious in our use of the word and we take care not to become the slaves of our terminology."[5]

If this is the character of all attempts to define "our" culture, then the question of whether other cultures have also had science must be solved on the same basis, and gives rise to the same problems, as the question of whether science existed in other periods of the culture which we conventionally consider to be our own. In every instance the answer depends on our acceptance of a given ideal of scientific knowledge; in other words, the problem we face when we attempt to understand science from other historical periods is analogical to the anthropological problem of understanding other cultures.

For these reasons, I will assume here that *the history of science constitutes the realization of a certain series of socially accepted ideals of science* which, though genetically linked, were distinct from one another; and that any attempt to decide at what point in this development we can say that we "really" have to do with science is an attempt to treat such an accepted ideal of science as if it were supra-historical, invariant and unproblematic.

An opposite position, though equally legitimate, would be to assume that science is the realization of only one of these ideals: that it did not exist until this ideal made its appearance, and that it will cease to exist when this ideal dies or changes. However, I do not know of anyone who supports such a view; those who treat science as a realization of one clearly defined ideal of science generally assume that this ideal is supra-historical.

The view accepted here eliminates the so-called problem of demarcating science from non-science or pseudo-science on the basis of normatively postulated methodological rules identified with criteria of rationality. It eliminates this problem because these criteria are understood as derived from accepted ideals of scientific knowledge. The accepted ideal of science defines not only the way in which the history of science is to be reconstructed, but also—and most importantly—the manner in which science is to be practiced in a given historical period.

2.

The concept of the ideal of science as understood here consists of a set of views about the goals of scientific activity and of views defining both the method and the ethos of science at a given period.

As opposed to the concept of the paradigm introduced by Kuhn, the concept of the ideal of science defines the criteria which determine what scientific paradigms of distinct research areas will be considered scientific at a given time, and enables us to speak of science as a synchronic whole, and not just as a collection of separate disciplines. More precisely, scientific activity oriented towards a given paradigm would be considered scientific only when its paradigm is consistent with the accepted ideal of science and constitutes its specification or elaboration for a given area of investigation.[6] The introduction of this concept into the theory of scientific development allows us, in my opinion, to avoid at least some of the difficulties encountered by the theory presented in *The Structure of Scientific Revolutions* and by its various continuations.

In the first place, it eliminates the difficulty pointed out by Feyerabend, who noted that even if every scientific activity is governed by a paradigm, not every activity governed by a paradigm is considered scientific.[7] In other words, it is impossible to define scientific activity without specifying its goal.

Secondly, we can eliminate Kuhn's assertion, which is rejected by virtually everyone, that the developed scientific disciplines are at any given moment in time governed by only a single paradigm. In accepting the apparently correct view that paradigms play a central role in the conduct of research in a given discipline or in the education of future scientists, there is no reason to exclude the possibility that a variety of paradigms or research programmes in a particular discipline can emerge on the basis of the same ideal of science. This seems to reflect far more adequately the actual history of science, in which the acceptance of a single disciplinary paradigm—raised by Kuhn to the status of a principle—has in fact appeared to be rather the exception (suggested by the history of astronomy) than the rule.

Thirdly, this conception allows us to differentiate clearly between a "local revolution," that is, a change of paradigm in some specific discipline or specialty, and a "global revolution" as a result of which the ideal of science accepted up to that point undergoes a change such that the goals and methods in the various disciplines are transformed.[8]

For example, the scientific revolution of the sixteenth and seventeenth centuries, as I shall argue below, consisted not only and not even primarily of the collection of discoveries which revolutionized the paradigms of astronomy (Copernicus, Kepler, Galileo), mechanics (Galileo and Newton), or chemistry (Boyle, Lavoisier), but rather in the formation and institutionalization of a brand-new scientific ideal which made these individual discoveries possible, and which was fundamentally different from the ideals which had guided the cognitive activities of scholars in antiquity and the Middle Ages. The scientific revolutions investigated by Kuhn occurred in individual disciplines within the framework of a particular ideal of science, and not as changes of the ideal of science itself, affecting the entire realm of scientific activity. As a result, for Kuhn, science does not constitute a whole, but rather a collection of various disciplines each defined by its own paradigm.[9]

Fourthly, if changes of disciplinary paradigms can occur without bringing about changes in the ideal of science (as, for example, in the rejection of Cartesian in favor of Newtonian physics), then the disruption of historical continuity in science, or the total collapse of consensus within a discipline, occurs far less frequently than Kuhn, and especially Feyerabend, suggest.[10] Moreover, given this view, one is no longer justified in maintaining the thesis that a scientific revolution completely destroys all possibility of communication between those scientists who accept the old and those who adopt the new paradigm, and that the transition from the one to the other is more in the nature of conversion to a new faith, and more the result of persuasion than of rational argument. The accepted ideal of science constitutes precisely this *consensus omnium* which makes possible the conduct of a rational discussion while a transition from one paradigm to another is taking place.

This does not mean that I believe that the introduction of the concept of ideals of science automatically eliminates all problems in the theory of scientific development. On the contrary, some of these problems, and especially this last one regarding the continuity and rationality of scientific development, reappear when one tries to explain the mechanisms of transition from one ideal to another. It is important to realize, however, that while changes understood as transitions from one disciplinary paradigm to another (local revolutions) are fairly frequent in science (and their frequency appears greater the more we restrict our conception of a specialty), changes in the ideals of science (global revolutions) are rare. By the same token, the difficult questions appear chiefly when we try to construct theories of scientific development *à longue durée*, while theories dealing with developments within a given ideal of science appear to be relatively free of these problems. Understood in terms of the concept of ideals of science, the theoretical image of science is not as paradoxically different from the traditionally accepted view of these matters as the views of Kuhn, and especially Feyerabend, suggest.

And fifthly, the introduction of this concept into the realm of theoretical reflection on the development of science alleviates, if it does not eliminate, the programmatic divergence of epistemology—together with the methodology of science—from sociology of knowledge. If one ignores the fact that the development of knowledge is co-determined by socially accepted and changing ideals, then sociological reflection on scientific research can easily turn into a simplistic explanation of this process as an unmediated result of economic or political influences on the content and direction of scientific development. This view denies the process of scientific development almost all its autonomy and its special internal logic; in the extreme version, the development of science is treated as a direct consequence of socio-economic changes. (This issue is discussed further in the Appendix.)

On the other hand, resistance to such a vulgar understanding of the mechanisms at work in the development of scientific knowledge usually means that epistemology is forced to treat cognition and its development as a result of immutable capacities of human nature, as an embodiment of its rationality—always the same, and always based on the same structures. The history of science is then presented as a completely autonomous "history of reason," and explained in the immanent categories of an ahistorical "logic of development," or "rational method," or even in the categories of the development of an autonomous world of ideas and problems—"world three," as Karl Popper has named it.[11] Whatever fails to comply with this logic is then considered as the result of extra-rational factors, deviance, or pathology. And as a result, there is no common ground between philosophy and sociology of science: the former investigates the rational mechanism of the development of science, and the latter the influence of extra-rational or non-rational circumstances deforming this development. The "internal logic" of scientific development is thus separated once and for all from its "external history"; and the only thing left to do is to specify or adjust the boundaries between them.[12]

I believe that the concept of the ideal of science can be helpful in bringing together these two approaches to the study of the development of science. This does not mean that the concept allows us to synthesize or overcome or eliminate the differences between studies of the "context of justification" and studies of the "context of discovery," between questions of *quid juris?* and *quid facti?* This is completely impossible. We can, however, study the process of scientific investigation based on an accepted ideal of science as autonomous, in the sense that it proceeds according to its own internal logic; while at the same time, we can understand this logic not as "a fact of nature," as an immanent expression of an unchanging "scientific reason," but as a "cultural fact" conditioned by the accepted ideal of science. This ideal is a historical fact rather than a "necessity of reason" and explains, for example, why certain rules for justifying claims will be considered rational or satisfactory while others will not. In short, the methodological rules governing scientific activity which constitute the basis for deciding questions of *quid juris?* are historically contingent and must cease to be treated as if they were expressions of the only possible rationality, immanent to human nature or to world three. They cannot serve as a basis for the formulation of demarcation criteria which would enable us to distinguish science from non-science once and for all. Nevertheless, they maintain their jurisdiction on the grounds of an accepted ideal of science.

As a result, both the ideal of "scientific reason" (the accepted ideal of science) and the means of realizing this ideal (the rules for pursuing science on the basis of this ideal) might become subject to philosophical critique. (It is usually only the latter problem which is raised by philosophical critique, while the ideal itself is treated as obvious, as the only possible one, and therefore unproblematic.) Sometimes an ideal of science which subordinates the cognitive function of knowledge to its technological or utilitarian functions might be deemed irrational if—in the view of the critic—its realization were to threaten certain positive cultural values. On other occasions, the critique might be directed against specific investigative procedures considered inappropriate or ineffective as a means of realizing an accepted ideal of science. The problem of the rationality of science and its development appears then in both spheres of the critique, but it does so differently.

Stanisław Ossowski claimed that the history of science, just like the history of any cultural sphere, depends "on what one thinks about it."[13] I find this formulation doubly appropriate: it is correct whether history is understood in terms of the real course of events, which is what Ossowski primarily intended, or when it is understood as written history, the historiography of science, as discussed above. A reconstruction of the ideal of scientific knowledge accepted at a given time would provide an occasion for both an investigation of its actual history and for a critical analysis of its written history.

Finally, it is important to emphasize that I do not identify the concept of the ideal of science with any particular philosophical or methodological positions, although there are links between them, as we shall see. The same proposed or accepted ideal of science can find its legitimation in various philosophical conceptions; and conversely, various philosophical conceptions can for different reasons deny it such legitimation. I think, for example, that Plato and Aristotle articulated the same ideal of science despite all the differences between them, and that the same ideal of science—though one very different from that of the ancient world—was defended by Bacon and Descartes.

It seems unnecessary to justify the claim that contemporary philosophical conceptions can articulate quite different ideals of scientific knowledge, or serve to defend or criticize the dominant, socially accepted ideal. Disputes between supporters of such different systems are generally disputes about the basic cultural values which science is expected to help to realize, or which—at the very least—science is not supposed to destroy.

3.

Let us consider some of the functions which the ideal of scientific knowledge performs in actual cognitive activity.

First, *such ideals demarcate the potential boundaries of the phenomenon called science*, that is, they determine what knowledge can be considered scientific, distinguishing those problems which belong to science and can be solved with its methods from others which go beyond its sphere of competence.

The ancient Greek ideal of science as an epistemologically certain form of knowledge (*episteme*) meant that technological knowledge was not considered scientific, since by its very nature it could be neither exact nor certain. As a result, Greek science had no applied disciplines.[14]

Exact technical knowledge is of course impossible without mathematical physics. But why did the ancients have no mathematical physics if they had mathematical astronomy? (Strictly speaking, the mathematical physics of Archimedes was forgotten until the Renaissance, when the conception of mathematical physics was first developed.)

As Alexandre Koyré convincingly demonstrates, the ancients did not create mathematical physics because they considered it simply impossible. Whether they followed Plato in believing that mathematical constructions have a real existence in the world of ideas and that mathematics is real knowledge, or whether they followed Aristotle in believing that such constructs have only a conceptual character and that mathematics as a field of knowledge is of secondary importance, Greek thinkers shared a common idea. They believed that "physical reality and mathematics are separated by an abyss," that "in nature there are no true circles, ellipses, or straight lines," and thus that the application of mathematics and measurement to terrestrial phenomena in order to search for certain knowledge does not make sense. This is why they did not admit that "precision could apply to this world, to terrestrial matter, to our own world, or that the sublunar world could embody mathematical being (unless it was forced to do so by art)"; accordingly, mathematical physics was impossible as an *episteme*.[15]

On the basis of the ancient ideal of science there was no place for the application of measuring apparatus (outside of astronomy), although some such instruments were used in everyday life (for example in the measurement of land or in weighing) when epistemological certainty was not at issue. In order to create mathematical physics, and in order to accept as scientific the knowledge based on such physics, it is necessary to believe that "the book of nature is written in the language of straight lines, circles and triangles" (Galileo); or in other words, that knowledge of the sublunar world can be accorded the same certainty as knowledge of the heavens. Such a conviction is necessary to justify the use of instruments in terrestrial physics. One could say that antiquity did not and could not give birth to Galileo (but could give birth to Copernicus), and that the use of measuring instruments as a means of scientific investigation required a fundamental change in the accepted image of the world, one without which Galileo's ideal of scientific knowledge could not be born. Only on the basis of this ideal did it become reasonable to use mathematics outside the realm of astronomy and to treat applied disciplines as a part of science.

The emergence of an ideal of science linking the cognitive function of knowledge with its technological function was made possible by this transformation: the emergence of an ideal thanks to which technical knowledge based on mathematical physics could be considered scientific. Yet this connection between the cognitive and the technical functions of knowledge is a historical fact rather than a necessity of reason. Its status as a value governing the development of scientific activity is a specific cultural phenomenon which could happen, but did not have to happen, since it does not derive from any logical necessity. This ideal of science is a product of the culture formed in Europe in the sixteenth and seventeenth centuries, which determined the direction of the development of modern science.

In chapters III and IV, we shall examine how this ideal was created, how it functioned, and what has led to its recent disintegration. As we shall see, this ideal not only broadened the potential sphere of scientific knowledge, but also excluded from it much that had belonged to it before —for example, philosophy. Let us note that the acceptance of the historical fact of the emergence of the modern ideal of science as if it were a "necessity of reason" is common both to those who regard the connection between the technological and the cognitive functions as a distinctive feature of human rationality, and to those who, for the very same reason, deny that science is rational and see it as a threat to culture. For both groups this ideal has a supra-historical character and is treated as obvious and unproblematic.

Secondly, *the ideals of scientific knowledge constitute a filter which determines that some research problems will be seen as available for investigation at a given time and are thus classified as worthwhile, interesting or important, while others will either pass unnoticed or be ignored as unimportant or even unscientific.*

The repertoire of questions which we can formulate is always conditioned by our present state of knowledge, by conscious and unconscious assumptions which we accept; the availability of some knowledge about a particular object is a precondition for the articulation of a question about this object. The actual state of knowledge about an object, and all possible questions which can be formulated about it can, following Popper, be called "the problem situation." I see no reason, however, for claiming, as he does, that the articulation of questions which can be formulated in a given problem situation exclusively concerns the logic of scientific development understood either as a method of scientific investigation or as a mechanism of the development of world three, the world of ideas and problems. Today, for example, "thousands of physicists and mathematicians use giant accelerators to investigate, at enormous expense, the behavior and characteristics of particles whose life-span does not exceed a millionth part of a second. At the same time, thousands of minor physical facts known from everyday life do not constitute a subject of interest to science. But after all, we have no way of knowing *a priori* that the mathematical patterns of the formation of foam on a beer mug or the path of a falling leaf might not prove to be cognitively just as interesting."[16] The history of science records an abundance of questions which were once formulated and then abandoned, and not necessarily because a solution was impossible at the time, only to be taken up again perhaps centuries later. And it is not uncommon to discover that many problems being formulated and addressed today could have been—in terms of the problem situation—addressed earlier.

All this means that the logic of the problem situation specifies only the range of questions that can be posed, and delimits a number of possible directions for the development of knowledge, without determining finally which of these questions will be formulated or which problems addressed and solved. This is why, as against Popper, I do not believe that the logic of development of world three could constitute even a good approximation of the mechanism of the development of science, or provide a model of its rationality such that on this basis one could reconstruct the development of science as a fully autonomous process relative to world one (the world of material reality) or world two (the world of the psychological experience of people and of the beliefs they hold). The realization of such immanent possibilities emerging from problem situations also depends in part on what specific ideal of scientific knowledge is actually steering the research conducted at the time. This is why any "rational reconstruction" of the history of science which fails to take into account how such ideals change and how they influence scientific research cannot give an adequate account of the real course of historical development and cannot constitute an adequate model for its evaluation as rational or irrational. This is so because the criteria of rationality on the basis of which such an evaluation is being made have not been derived from an historical analysis but are posited *a priori*.

Thirdly, *ideals of scientific knowledge (together with accepted ontological and epistemological beliefs) co-determine the rules governing the acceptance and rejection of claims, the principles of adequate explanation of phenomena, and methods of constructing theories. In other words, ideals of science co-determine the methodological rules of research.*

Since this thesis is of general importance for the whole of this work, I shall try to examine it in detail in the following chapter on the basis of disputes about the rules of explanation in science. Here I shall limit myself to a few general remarks.

If methodological rules are not historically invariant but instead depend on an accepted ideal of science, then no methodology can constitute a supra-historical means of distinguishing science from other fields of intellectual activity or from its products. This is why, as I have already indicated, all criteria of demarcation based on methodological rules turn out to be historically inadequate, since they do not encompass and cannot encompass everything that historically has been considered science or that possibly might yet be considered science. This is so because all definitions of science based on methodological criteria are based on the assumption that the criteria of scientificity are invariant, that they can be discovered once and for all, and that they constitute an expression of human rationality. De facto, what such definitions allow us to distinguish at best is what can be considered science according to the currently accepted ideal; they allow us to construct theories of scientific development in the short term (although here too they encounter serious problems, which we shall address in chapter VII).

> This view, according to which methodology is an empirical science in its turn—a study of the actual behavior of scientists, or of the actual procedure of 'science'—may be described as '*naturalistic*'. [...] Thus I reject the naturalistic view. It is uncritical. Its upholders fail to notice that whenever they believe themselves to have discovered a fact, they

have only proposed a convention. Hence the convention is liable to turn into a dogma.[17]

However, the author of *The Logic of Scientific Discovery* goes on to say that "It is only from the consequences of my definition of empirical science, and from the methodological decisions which depend upon this definition, that the scientist will be able to see how far it conforms to his intuitive idea of the goal of his endeavours."[18]

While agreeing with Popper's critique of the naturalist position, we have to take note of the fact that such an intuitive idea of the goal of investigations, or the socially accepted ideal of science, as I prefer to say (for otherwise, where does the scientist draw his intuitive idea from, and why do most scientists share the same basic intuitive idea at a given time?), is not historically invariant. Accordingly, whenever we treat methodological rules and the criteria based on these rules as universally valid, we absolutize the ideal from which they are derived and we cease to perceive the fact that this ideal (or the intuitive idea of a goal) is neither eternal nor the only one possible. We cease to treat it as a historical fact, the result of a particular stage of cultural development, and we present it as a necessity of reason. Such a formulation conceals an evaluative judgment of the definite (but by no means the only possible) form which science assumes currently. As a result, the problem of the rationality of knowledge, which was to be guaranteed by the method, is reduced to a rationality now understood only as a rationality of the means (methods) serving to realize a socially accepted ideal of science. The question of the rationality of this ideal itself, with respect to the cultural values which science is expected to serve, no longer appears at all; it was answered *a priori* in the form of its unproblematic acceptance. All alternative possibilities remain beyond the horizon of the philosophical reflection on science conducted in this manner.

Despite the indubitable differences between statements concerning the subjects of investigations which together form the current state of knowledge, and normative methodological rules, both the former and the latter must be treated as elements of a single system. And changes in the methodological rules demand explanations to the same extent as changes in the content of the scientific theories which we construct and accept using these rules. The history of methodology seen from this point of view remains an unwritten chapter in the history of science, one which would be valuable if only because it would constrain the tendency to treat the current methodological rules and the actual state of science as the only possible and rational ones.

Although, as I already mentioned, methodological controversies often result from the acceptance of different ideals of science, various methodologies can nevertheless be formulated on the basis of the same ideal, depending on the accepted ontological and epistemological beliefs. In other words, *the accepted ideal of science allows for some methodological conceptions while excluding others, but does not impose any such views apodictically.* If, for example, science consists of knowledge which is certain, then it cannot include statistical predictions or probabilistic explanations of phenomena; but, obviously, there might be many different ways of achieving certain knowledge. It is also difficult to imagine how instrumentalism or operationalism could have emerged in the context of the ancient ideal of science. A methodology which is devoted to the fastest possible revolutionary changes of knowledge, and to the introduction of theoretical innovations, surely could not have been born within the framework of the ideal of knowledge common in the Middle Ages, when constancy rather than change served as the main value of cognition. A scholastic method of studying texts probably accords with this ideal just as closely as Popper's falsificationism fits the contemporary ideal.

It is in any event important to emphasize that methodological conceptions originate not only in an accepted ideal of science but also in the prevailing ontological and epistemological context. This enables us to understand how it is possible for methodological controversies to occur despite the acceptance of a common ideal of science. Contemporary disputes between logical empiricism and Popperian falsificationism can serve to illustrate this situation. Incidentally, Popper himself stresses the similarity of his views to those of Carnap or Reichenbach in one of the footnotes in *Conjectures and Refutations*, while at the same time he conducts a polemic on virtually every methodological issue. The constant disputes as to whether Popper is or is not a neo-positivist, which one finds so often in the literature of our field, have their source, I believe, precisely in the fact that no distinction is drawn between methodological positions and accepted ideals of science. In the first respect, Popper is certainly not a neo-positivist; while in the second—as we shall argue—the differences between him and the logical empiricists indeed appear to be minimal.

Fourthly, *ideals of science imply a particular scientific ethos and internal organization of the scientific community, as well as their understanding of science as a social institution.*

It is often said that truth is a value to which both methodological rules and the ethics of scientists are subordinated. But when it happens that truth in science is valued at one time as an autonomous value requiring no further justification, and at another time as an instrumental value—that is, precisely as a means of realizing other values (economic or political power)—then it seems to me that we are justified in concluding that in the two cases we are dealing with different social institutions which have different social organizations and occupy different places in the global social structure, and whose members have different kinds of ethos and different forms of cooperation and competition. Such differences in the evaluation of true knowledge will have an immediate influence in deciding issues such as the autonomy of science *vis-à-vis* other spheres of social life, the freedom to communicate results, disinterestedness, the moral responsibility of scientists for the results of their work, and even the evaluation of scientific achievement.

The assessment of true knowledge also depends on the addressee of scientific work. Sometimes competent colleagues and peers, who regard achievements as the basis of the formulation of new research problems, constitute the primary, though not necessarily the only, audience; while on other occasions it is mainly people and institutions outside science for whom such achievements are of interest, especially in view of the possibilities of their practical utilization for the manipulation of objects, symbols, and people. Communities of scientists aiming at one or the other of these groups cannot share a common ethos, cannot have the same internal organization or possess a similar kind of self-knowledge. Their relations with the outside world must also be different. A scientist who would like to work towards both of these goals simultaneously would resemble an actor trying to perform in two different plays on two stages simultaneously in front of audiences with very different artistic tastes. The great majority of contemporary scientists are required to take part in both of these plays, and often they are not even aware of how very different they are.

Situations of this sort lead to a variety of moral conflicts. As a result, a scientist can no longer believe that his moral responsibility as a scientist is limited simply to his following the methodological rules of the game. This forces him to reflect on his own activity and his social status in a manner different from that to which he was previously accustomed. What is today often called the "crisis of science" seems to be rather a crisis of a particular ideal of science which has ceased to fit reality and whose defense is more and more often a result of false consciousness. This is the crisis which gives rise to disputes about the rationality of science and of its development.

I think that these initial remarks concerning the ideals of science indicate the theoretical context in which we have to examine the disputes about the rationality of science, which is our central concern. First, however, I shall try to elaborate the view outlined above on the basis of a controversy concerning the rules governing scientific explanation, in order to examine in detail how the ideals of science co-determine the methodological rules of its conduct.

CHAPTER II

IDEALS OF SCIENCE AND RULES OF EXPLANATION

1.

The term *explanation*, as commonly used, designates a variety of procedures whose goal is to bridge the gaps in our knowledge, both in matters of everyday life and when scientific issues are at stake. We ask for an explanation when we do not know how to do something, or why a certain event has occurred, or why there is a certain regularity in phenomena, or the meaning of something we treat as a sign. We explain to a child how to multiply fractions or how to play chess. We explain human actions by appealing to their physical causes and/or psychological motivations. We explain why there are eclipses of the sun, why a war has been declared, or why bodies fall at a constant rate of acceleration. We explain the meaning of a novel, or a custom, or a dream.

In everyday language we are rarely able to establish unequivocally the meaning of a word in all its various uses, and this is also the case with the term *explanation*. The statement with which I began, to the effect that the goal of an explanation is to bridge gaps in our knowledge, certainly does not constitute such an unequivocal definition, at least not until we can specify what kind of knowledge is to be supplied by an explanation. The term *knowledge* is after all no more clear than the term *explanation*. In this difficult situation it would of course be possible to behave like the authors of dictionary entries: to explain the meaning of the term with reference to its various uses, and thereby to avoid the question of whether it has a single meaning common to all these usages. But this is not where the problem lies. What is at issue after all is not a description of the various uses of the term *explanation*, but a solution to a methodological problem which could be formulated as follows: what conditions must be met by an explanation so that it really can bridge the gaps in our knowledge? In other words, what we want to ask is: *what constitutes a scientifically adequate explanation?*

Although this question has been posed since antiquity, none of the answers proposed have achieved general approval. Thus we might begin by inquiring about the nature of the problems encountered when one tries to answer it.

27

There is no doubt that there are many logical, epistemological and pragmatic difficulties involved. There have been disputes about the logical character of statements used to explain (i.e. of the *explanans*), about the cognitive conditions that must be met before an explanation can be accepted, about the logical connections which must obtain between the *explanans* and the *explanandum* (the statement reporting what needs to be explained), and about the material relations which must be established between what is asserted by the *explanans* and the *explanandum*.

Without trying to minimize all these problems, to which a great deal of literature has already been devoted, I believe that in a certain sense they are of a secondary character, and that the primary difficulty, which must be resolved before the other problems can be addressed, resides elsewhere. Let me illustrate this initially with two examples.

Some contemporary language philosophers believe that an analysis of the everyday uses of terms (and thus also of the term *explanation*) can and should lead to the discovery of their actual meanings, and by the same token enable us to distinguish between adequate and inadequate explanations. Other philosophers argue that the everyday usage of terms cannot reveal their real meanings because language is not only a means of communication but also of describing the world. For this reason, analysis of the everyday usage of a term (such as *explanation*) in communicative situations cannot reveal what an explanation in science should be, because language functions here primarily as a means of description. So while common language philosophers claim that scientific explanation should be modelled on whatever passes as an adequate explanation in everyday life, their opponents claim to the contrary that it is precisely the scientific model of an explanation that sets the pattern for explanations which can be considered universally adequate.

Now to the second example: Popper writes that "it is the aim of science to find *satisfactory explanations* of whatever strikes us as being in need of explanation."[1] While agreeing with this formulation I must nevertheless ask: what determines whether or not a given explanation will be considered *satisfactory*? For example, does it have to be formulated in categories referring to observable phenomena, or can it refer to objects, characteristics or relations which are not open to direct observation? Is it a question of explaining the unknown by what is directly known, as Aristotle postulated, or of explaining the known by what is unknown and hidden, that is, of deepening our knowledge of the world—as Popper argues? Must the explanation that is given be technically operational, allowing for the reproduction or prediction of the explained events, or does it have to provide an intuition of understanding? Many authors appear not to notice that every answer to these and similar questions, i.e., every acceptance of a specific model of scientific explanation, implies the acceptance of particular values whose realization is to be assured by the development of knowledge.

Both examples suggest that the manner in which the criteria of satisfactory explanation are formulated will depend on the accepted ideal of scientific knowledge. When Popper claims that the aim of science is to provide satisfactory explanations, yet ignores the question of why the model he proposes should be considered "satisfactory," it is apparent that he considers the ideal of science he has accepted to be unproblematic and universally valid.

Popper does of course list the criteria which every explanation must satisfy in order to conform to his normative view of science; yet if we want to avoid circularity, we cannot answer the question of "why this model of explanation is satisfactory" with the statement "because it conforms to my conception of science," since it is precisely this conception which is at issue. Only when a definite ideal of science has been accepted can we then—appealing to it—try to resolve the problem of what criteria of adequate explanation are best suited to it. In other words, only then is it possible to deal with the aforementioned logical, epistemological and pragmatic issues—which, however, could assume a completely different formulation, or not appear at all, within the framework of some other ideal of science. If the methodologist treats the currently accepted ideal of science as unproblematic, then only these kinds of issues remain to be resolved.

I would risk the claim that it is here that we can draw a tentative boundary between doing philosophy of science and being concerned with scientific methodology. This boundary marks the difference between an expert who is expected to indicate the means required to achieve a goal that was chosen earlier without his participation, and an advisor participating in the formulation of goals.

Disputes about the model of adequate or rational scientific explanation can serve us here as examples in our attempt to justify the thesis that methodological rules depend on the accepted ideal of science. In order to avoid historical digressions, I shall limit myself to trying to demonstrate that at least some contemporary disputes about criteria of adequate explanation are basically derivatives of the controversy about the ideals of scientific knowledge. I will not, however, try to address the issue of how the rules of explanation have changed historically together with changes in the accepted ideals of science. I will begin with a discussion of the so-called "covering-law model" of Hempel and Popper, since this has been the focus of numerous controversies about explanation in contemporary literature.

2.

The main idea of the Hempel-Popper model is the claim that the *explanans* of every satisfactory scientific explanation must contain at least one general statement (law) which has either an exceptionless or a statistical character and must logically imply or inductively justify the *explanandum*.[2] Accordingly, the model distinguishes among three types of law-like explanations which differ in their logical structure:

a) nomologico-deductive explanations;
b) nomologico-statistical explanations; and
c) inductive-probabilistic explanations.

In the first two instances the *explanandum* follows by deduction from the *explanans*, which contains either (a) an exceptionless law or (b) a statistical one; in the third case (c), the *explanans* does not imply the *explanandum* but justifies it inductively.

I shall forego here a discussion of the logical and epistemological problems connected with the elaboration of this model; they are well known and have been discussed exhaustively by the authors of the model. I shall also ignore the often essential differences among the advocates of this model, which concern specific issues and result from differences in their ontological or epistemological positions. I shall focus instead on critiques which either question generally the adequacy of this model for the explanatory procedures used in science, or those—the most common ones —which question its universal validity.

The first critique which needs to be noted is that explanations of this kind do not explain any phenomenon fully, in all its detail and all its complexity, but rather treat it as an element in an abstract class of events of a particular type. For example, when we ask for an explanation of everyday events or of historical phenomena, we treat these events as concrete and unique, and not as instances of such abstract classes of events. If we ask a doctor to explain why a patient died, we will not be satisfied with the answer: "because he was a man and men are mortal," although it is obvious that this statement fulfills the requirements of a correct nomological explanation. Similarly, when we ask for an explanation of a concrete historical event, we might not be satisfied with an answer which explains the occurrence of a certain type of event of which the event in question was one instance, even if it does so correctly, since such questions are not asked in order to find out why revolutions, strikes or wars happen in general, but why a specific war, a particular revolution, or a given strike actually occurred. A doctor, or a historian, can of course make his answer more elaborate, as usually happens in such situations; but eventually it will always turn out that an explanation by law treats a given case as an element of a class of analogous events and fails to explain some of its individual characteristics which distinguish it from other similar events. Regardless of the amount of detail accorded a description of this class, explanation by law will always take the form: "under certain circumstances certain events always (or usually) take place, and these circumstances are present in this case." If all the characteristics of the event to be explained could be described fully, then explanation by law would be impossible, since laws are statements referring only to classes of events. The process of formulating a description of the state of affairs in question will eventually lead to a situation in which the *explanans* no longer contains a general statement (law). Let us note at the same time that a consistent demand for an individualizing description of the event to be explained cannot stop with the abandonment of all reference to general statements (laws). If such a demand were to be consistently formulated and implemented, it would in effect prohibit us from using general concepts.[3]

And so, in seeking to explain individual, concrete events we face a dilemma: either our explanation is incomplete (does not explain all the characteristics of the event to be explained), or it fails to meet the conditions imposed by the nomological model of explanation. It follows from this that explanation by law can be treated as a model of explanatory procedure only under the tacitly or explicitly adopted assumptions that (a) the knowledge which an explanation can provide must have a general, nomological character, and (b) knowledge which does not contain laws is not scientific. However, this conception of scientific knowledge is not generally accepted, especially by social scientists and students of the humanities who believe that, as opposed to the natural sciences, the humanities and the social sciences do not pursue the formulation of regularities which would allow us to predict events and manipulate objects, but seek rather to understand concrete events and processes and explore their human significance. Such an understanding cannot be provided by explanations which do not treat the events and processes as fully concrete. Thus the argument that the model of explanation by law is unsatisfactory, since it cannot explain events in their particularity, is based on an ideal of scientific knowledge that is different from the ideal on the basis of which the covering-law model was developed. The situation is similar in the case of other arguments directed against this model.

3.

In an article criticizing the nomologico-deductive model of explanation, Michael Scriven claims that this model "leaves out of account three notions that are in fact essential for an account of scientific explanation: context, judgment and understanding."[4] By the same token, the nomological model cannot provide an adequate account of what the explanatory procedures should be either in science or in everyday life. The author claims also that the real meaning of the term *explanation*, common to all its uses, can be discovered through the study of the use of explanatory procedures in everyday life. Thus it is not science but everyday life that is to provide us with the common model we are seeking.

The controversy is apparent: one side claims that there is no explanation without appeal to general statements, and that this manner of explaining is a model for all explanatory procedures; while the other side maintains that "it is the *understanding* which is the essential part of an explanation,"[5] and that one can usually achieve such understanding without recourse to laws (which might serve to justify explanation but are not a necessary element), and that this type of explanation can serve as a universally valid model.

Let us now try to identify the crux of the controversy. We may begin with "context."

As Scriven writes,

> whatever an explanation actually does, in order to be called an explanation at all it must be capable of making clear something not previously clear, i.e. of increasing or producing understanding of something. The difference between explaining and 'merely' informing [...] does not, I shall argue, consist in explaining being something 'more than' or even something intrinsically different from informing or describing, but in its being the appropriate piece of informing or describing, the appropriateness being a matter of its relation to a particular context.[6]

He adds further:

> The primary case of explanation is the case of explaining *x* to someone [...] For it makes no sense to talk of an explanation which nobody

understands now, or has understood, or will, i.e., which is not an explanation for someone. In the primary case, the level of understanding is that of the person addressed. The notion of 'an (or the) explanation of *x*' makes sense just insofar as it makes sense to suppose a standardized context.[7]

It makes no sense to pose the question of what constitutes a satisfactory explanation without reference to its addressee. An explanation is satisfactory not when it fulfills some objective methodological criteria, but when it supplies psychological understanding: "[P]roblems of structural logic can only be solved by reference to concepts previously condemned by many logicians as 'psychological, not logical,' e.g., understanding, belief, and judgment."[8]

Let us note that once we consider the problem of context, there is a fundamental difference between scientific explanation and everyday explanation, regardless of what the concept of explanation refers to. When we demand an explanation in everyday life, we do not expect our interlocutor to provide information that was previously unknown to anyone. On the contrary, we want to learn something that we believe *to be known but that we do not know*. In other words, we assume the existence of some socially available knowledge which contains an answer to our question, and we ask to be given this answer. (Of course, in everyday life we could demand explanations which at that point in time are not known to anyone; but this possibility does not concern us here.)

Thus the procedure for requesting explanations is here obviously a case of interpersonal communication, and the problem of context cannot be ignored. It exists for both sides: for the person posing the question as well as the one answering it. The former assumes that the answer to his question is known, and moreover, that the person who is being asked knows it; while the latter assumes something about the knowledge of the person addressing him: sometimes he must guess what the question really is about or why an explanation is being requested, and he adjusts his answer (explanation) accordingly. In any case, the person asking the question certainly does not believe that the person he is addressing will tell him something that no one has ever known before then. In other words, a certain "inequality of knowledge" is presumed between the person asking the question and the one answering it.

When we request explanations in science, the situation is completely different. We are searching for an answer that nobody knows, which must be discovered as the result of a process of cognition, rather than simply being communicated. It is one thing to provide an answer which is generally known and acknowledged as scientific, and quite another to search for a solution to some new, as yet unresolved problem.

This does not mean that in this second case the issue of context does not arise at all; but it is of a completely different kind than in the case of interpersonal communication. Here one assumes a precise "equality of knowledge" between the person asking and the one answering (quite often they are the same person), and thus the existence of a "standardized context," to use Scriven's expression. In any case, it is tacitly assumed that the person asking the question knows everything there is to be known about the subject at the time, and that the answer to the question being posed cannot be obtained simply as a result of interpersonal communication, since no one knows it at this point. The issue of "to whom is the answer addressed?" is simply unimportant in this case, since it is assumed that anyone to whom it might be addressed will have access to precisely the same knowledge about the subject.

The problem of context in the case of scientific explanation is basically a result of the fact that the "standardized context" is historically changing. This is so not only because of changes in the stock of socially available knowledge, but also because this context is constituted by different conceptions of the methods and goals of scientific cognition— including models of adequate explanations. The fact that scientific explanation is also relativized to a particular historical context does not mean that this context should be seen as equivalent to the context of interpersonal communication, and it certainly does not mean that we should treat explanation in everyday life as a model for all adequate explanation. In order to grasp the source of this idea, however, we have to examine the issue of *understanding*.

The statement that the natural sciences do not allow us to "understand" phenomena was formulated already by Dilthey and was supported by a number of authors, especially in discussions concerning the methodological relations between the natural sciences on the one hand and the humanities and social sciences on the other. We shall return to this issue; here we need only note that there is an essential difference between the views of Scriven and those of Dilthey regarding their notions of what constitutes the "understanding" which explanations are supposed to provide, and that this difference is at the source of the idea that interpersonal communication can provide a model of "satisfactory explanation" for science.

Let us examine some of the examples of everyday explanatory processes which, in Scriven's view, lead to understanding. He argues that there are clearly cases when we can explain without language, e.g. when we explain to the mechanic in a Yugoslav garage what has gone wrong with the car. Now this is hardly a scientific explanation, but it seems reasonable to suppose that the scientific explanation represents a refinement on, rather than a totally different kind of entity from, the ordinary explanation. In our terms it is the *understanding* which is the essential part of an explanation . . .[9]

So, using only gestures, we can explain to a Yugoslav mechanic what is wrong with the car we want him to repair, and we know that he has "understood us" because he could repair the car. The teacher explains to the student how to multiply fractions, and the student has understood if he can correctly solve the problems in his book. We explain to the passer-by how to get to city hall, and he has understood if he managed to reach his destination. Had we told him that "the city hall is situated in the same relation to the post office as in all other towns," our (nomological!) explanation would not provide him with an "understanding," since he does not know what this relation is. If he did know it, he would not be asking for directions.

All these examples taken from Scriven's text show clearly what sort of explanation and what sort of understanding he has in mind, and demonstrate that this has nothing to do with the "understanding" of Dilthey or of other hermeneutic philosophers. Scriven does not appeal to any state of mind; he does not require empathy. The question of whether understanding was achieved or not is resolvable by intersubjective methods, that is, by observation of the behavior of the person who posed the question. This is obviously a behaviorist criterion of understanding. Was the explanation understood? In order to answer this question we must communicate with the person who asked it and observe his behavior. If he reacts in an adequate manner, he has achieved "understanding"; otherwise, we must continue our explanation.

In my opinion, this notion of understanding, together with the conviction that explanation should provide understanding, is at the source of the view that a model of explanation must be based on everyday explanatory procedures. It is only interpersonal communication that can supply an intersubjective method of testing whether understanding was achieved or not.

It is typical that even when Scriven talks of scientific explanation, his examples never refer to situations involving a *search* for unknown explanations, but only to the *communication* of accepted explanations. He writes for example,

> If you reach for a cigarette and in doing so knock over an ink bottle which then spills onto the floor, you are in an excellent position to explain to your wife how that stain appeared on the carpet, i.e., why the carpet is stained [...]. You knocked the ink bottle over. This is the explanation for the state of affairs in question, and there is no nonsense about it being in doubt because you cannot quote the laws that are involved [...]. Having reasons for causal claims thus does not always mean being able to quote laws.[10]

It is easy to see that this entire reasoning does not refer to a process of investigation, but to the presentation of known explanations. It is also obvious that such reasoning would be entirely useless if what was at stake was an explanation of an event which science at a given moment is unable to explain: for example, if we were unable to indicate any other event which could serve as a cause of the state of affairs to be explained. In order to state that one event is the cause of another, we must formulate an appropriate regularity.

Scriven is undoubtedly right when he says that in order to explain the spot on the carpet it is enough to point to a cause of this fact without explicitly formulating any laws. But he is right only because in this case the law in question is generally known. Either all causal explanation is nomological, or the meaning of the cause-effect relation cannot be reduced to the formula "always if A then B." How then, however, can we convince ourselves that A is indeed the cause of B?

The fact that we are aware of some regularity is not by any means equivalent to our ability to formulate a law. Certainly people have always known that heavy bodies fall; they certainly knew this before any theory of physics was formulated to explain it. If someone "understands" that the carpet was stained as the result of an ink bottle being knocked off a table, this does not mean that he reaches such an understanding without knowing the appropriate regularity. It is the context of everyday knowledge which decides that the two events will be seen as causally connected. Such a context exists only when we are able to communicate an already known explanation, and not when the explanation remains to be discovered. If we take into account the difference between these two situations, the claim that laws do not constitute a necessary part of the *explanans* but rather serve for its justification becomes nothing more than a verbal trick.

Still, the conviction that explanatory procedures should supply understanding can be articulated not only on the basis of a behaviorist conception of the human psyche and cognition.

4.

The validity of the nomological model is also questioned by those who claim that explanation by law is impossible in the social sciences and humanities, since these disciplines simply do not produce claims which might be considered universally valid (valid always and everywhere), and by those who believe that the formulation of such laws, even if it were possible, would still not constitute the main goal of these fields.

> One has a right [...] to ask those who assert that the aim of social anthropology is to formulate sociological laws similar to the laws formulated by natural scientists to produce formulations which resemble what are called laws in those sciences. Up to the present nothing even remotely resembling what are called laws in the natural sciences has been adduced.[11]

As Alan Donagan expresses it, "The existence of false sociological hypotheses cannot show that there are true historical explanations which rest on covering laws."[12]

Indeed, the current situation in the social sciences is such that the universal statements used to explain phenomena do not have the character of strictly universal statements; and if they are formulated without being made relative to a specific space or a specific historical period, they always turn out to be false. However, if they are formulated as specific to a given time and space, they cease to be "laws," at least in the sense in which the supporters of the nomological model use this term.

The issue of whether this state of affairs is accidental and temporary or whether it follows from the very nature of social knowledge is highly controversial. Some scholars claim that social knowledge is itself a "social variable" on which human actions depend. If so, then laws in the social sciences cannot abstract from such historical circumstances as social knowledge about the regularities of social development. The essential difference between knowledge of the natural and of the social world would then consist in the fact that the behavior of natural objects in no way depends on what people at a given time know about this object; while social processes depend in part on such knowledge, so that the regularities describing such processes cannot, if only for this reason, have a universal character. Learning about social processes might change their course.

Some of the theses of structural anthropology counter this argument by trying to show that social processes depend on certain universal structures, that is, on systems of formal relations which do not belong to the realm of phenomena of which social actors are conscious. If this were correct it would mean that it is possible to discover certain social regularities which would be supra-historical, transcultural, and just as independent of our knowledge as the regularities in the natural world.

Whatever the solution of this issue might turn out to be, it remains a fact that the explanatory procedures in the social sciences have not thus far been based on appeals to such timeless and universal regularities. For this reason alone, it would be difficult to insist that they correspond to the nomological model under the assumption that the status of laws can be attributed only to strictly universal statements. I have argued elsewhere that this assumption is itself debatable.[13]

I believe namely that the difference between the natural and the social sciences is in this respect not so radical, since the natural sciences also do not formulate strictly universal statements; or, more precisely: it is impossible to prove that statements considered natural laws actually do have the status of strictly universal statements. The decision about whether a given claim can be considered strictly universal (i.e. a law) depends on the entirety of the available knowledge. Statements which we consider as laws concern classes of objects which are not so much ontologically as epistemologically open; in other words, they are not statements derived from historical generalizations that could be formulated as a result of the listing of all such cases, even if the number of cases were known to be ontologically finite.

We can accept a nomological model of explanation while still not accepting the concept of law adopted by those who have formulated this model and by some of those who support it. If we do so, the argument that the nomological model is useless because social sciences formulate no laws ceases to be applicable.

Still, even if one accepts such a "liberalization" of the conditions which must be met before a statement can be considered a law, one could still claim that although a nomological explanation in the social sciences is possible, such explanations are still not satisfactory for other reasons. It is this view that most concerns me here.

What is common to all the arguments supporting this view is the conviction that as against the natural sciences, explanation in the social sciences and especially in history must be formulated in terms of the psychological motivations of the actors, and that the natural sciences do not have to deal with this problem because of the nature of the objects they investigate. Such motivations can be discovered only on the basis of the internal psychological experience of the investigator, who must be able to imagine himself acting in the situation which he is investigating, so that through introspection he can decide what sort of motivations might have led people to act as they did. An exposé of these motivations will supply him with the understanding pursued by the social sciences and humanities. Herbert Butterfield, for example, writes that

> the historian must put himself in the place of the historical personage, must feel his predicament, must think as though he were that man. Without this art not only is it impossible to tell the story correctly but it is impossible to interpret the very documents on which the reconstruction depends. Traditional historical writing emphasises the importance of sympathetic imagination for the purpose of getting inside human beings.[14]

In the face of such argumentation, a defender of the nomological model would claim that one must distinguish situations where an internal psychological state is the source of a specific explanatory hypothesis from situations where it serves as a justification of this hypothesis. Thus, he does not deny the fact that the sympathetic imagination can suggest a fertile hypothesis to the investigator and play an important heuristic role; he will claim, however, that the question of whether the hypothesis is satisfactory does not depend on how the historian arrived at its formulation, but on how it can be justified.[15] In other words, he will stress the difference between the rôle of statistical experience in the process of the discovery of the explanatory hypothesis and its rôle in the process of justification. And no matter how valuable this experience can be heuristically, by its very nature it cannot constitute an intersubjective justification of the proposed explanatory hypothesis. As Edgar Zilsel puts it: "When a city is bombed it is plausible that intimidation and defeatism of the population result. But it is plausible as well that the determination to resist increases. [...] Which process actually takes place cannot be decided by psychological empathy but by psychological observation only."[16]

This reasoning will not satisfy the polemicist. He will claim that internal experience is not just an heuristics but a basic methodological tool used for the interpretation of human behavior, regardless of whether or not this behavior is verbal. Social events are cultural facts, and as such they are not only causally conditioned but also have a sense: they mean something. To explain them we have to discover their meaning. In studying a given custom we want to know not only why people behave in this manner, but also what this behavior (or deviance from it) means in a given culture. Internal psychological experience is one method, or even perhaps the only method, which allows us to reveal this meaning: this is so first because everyone has psychological experiences in various situations; secondly, all human beings share a similar psychological structure; and thirdly, we are entangled in a cultural context thanks to which we can grasp the meaning of the phenomena which surround us. Thus, all the proponents of this position, from Dilthey to Ricoeur, have always advocated various psychological methods for the interpretation of social phenomena and treated them as hermeneutic methods for investigators of these phenomena. Nomological explanation perhaps allows for discoveries of causal conditioning, but the social sciences must interpret phenomena, and "interpreting," as Freud was wont to claim, "means finding hidden meanings."

A defender of a nomological model would say in turn: let's assume that this approach to social research is justified, and that social phenomena are indeed meaningful and that the aim of the social sciences is to discover these meanings. These are assumptions currently accepted in the humanities. This does not mean, however, that hermeneutics is the method by which discoveries should be made, and it certainly is not the only method leading to this goal. Such a conclusion demands the acceptance of certain assumptions in addition to those presented above.

We encounter here first of all an expressive theory of culture according to which social phenomena are the expression of hidden, psychologically understood meanings which can be discovered through the study of the conscious or unconscious motivations of actors. Without such a psychological formulation of the concept of meaning one would be unable to claim that hermeneutics of any sort provides an access to it. But this is by no means the only possible way to understand the "meaning" of a social fact. If, for example, phenomena are understood as meaningful signs fulfilling specific functions and occupying specific places in a system of signs, then their meaning can be discovered through the study of this system and its structures. The structure of the system can reveal the meaning of the phenomena; and no hermeneutic method can constitute a means of discovering the structure if the meaning of a sign, regardless of its nature, depends not on the consciousness or unconsciousness of its creator, but on the place that this sign occupies in a system.

In its contemporary form the above controversy is probably best illustrated by the polemics between Claude Lévi-Strauss and Paul Ricoeur about the structural interpretation of myths. It is easy to see that the crux of the controversy was a different understanding of what is meant by the "meaning" understood as the subject of research in the social sciences and humanities. Ricoeur states:

If the meaning which I grasp does not increase my understanding of myself or of things, it does not deserve to be called meaning. [...] What do we understand by the words *sense* or *nonsense* if not episodes in a consciousness of history, which is not simply the subjectivity of one culture looking at another culture, but truly a stage in the reflection which is trying to understand everything? In other words, what is meaningful are particular discourses, what has been said, and not simply their syntactic arrangements viewed by an external observer. I mean that in order to do science we must limit ourselves to the consideration of arrangements of which one is oneself an observer. Thus one avoids entering into what I call the 'hermeneutic circle,' which makes me into one of the historical segments of a content which interprets itself through me; [...] Can one still speak of sense and nonsense if this sense is not an episode in a fundamental reflection or a fundamental ontology [...]?[17]

Lévi-Strauss answers him as follows:

It seems to me that you are connecting the concept of discourse with the concept of a person. What do the myths of a society consist of? They constitute the discourse of this society, a discourse which does not have a personal sender: a discourse, therefore, to be examined as a linguist studies a little-known language, trying to unravel its grammar without regard for who said what was said. [...] What do I understand by meaning? It is a specific aroma caught by consciousness when it tastes a combination of elements none of which taken separately would exude such an aroma.[18]

And so, one can oppose to the expressive theory of culture its systemic formulation, in an approach according to which the meaning of a sign is defined by its place in a system of relations which can be grasped from the outside, objectively, through the investigation of the structure of such a system. The fact that social phenomena are meaningful by no means implies that these meanings have to be interpreted psychologically or that they can only be revealed by means of hermeneutic methods. It is true, however, that the interpretation of social phenomena is impossible without some grasp of their meaning; but such an admission does not necessarily imply either that we must reveal the psychological motives of actors, or that we enter a hermeneutic circle which makes the interpreter himself into one of the historical segments of a content finding its interpretation through him. Revealing structural regularities allows for an objective interpretation of signs and of their meanings.

5.

I will close my account of the dispute here, but not because I believe the arguments I cited have settled the issue once and for all. The entire argument was cited here not so much in order to solve the problem of what would be satisfactory rules of explanation, as rather to demonstrate their dependence on an accepted ideal of science. This was the reason for ignoring some important disagreements both among the supporters of the nomological model and among its critics. I interrupt my account of this controversy because I do not believe that the conflict presented here can be settled either philosophically or methodologically.

It is unresolvable philosophically because when we place ourselves outside the system which we are investigating and of which we are a part, we cannot achieve full knowledge of that system. If we accept that we are a part of this system, we cannot achieve objective knowledge of it. The conflict between our striving to describe a system from the outside and our understanding of ourselves as a part of this system is unresolvable. In striving for both of these goals, we are constantly torn between two perspectives. What defines our choices are the values we seek to realize through the knowledge achieved. The acceptance of a particular ideal of science depends on the choice of these values.

Methodologically, the conflict is unresolvable precisely because there are various ideals of science in the background which preclude the acceptance of certain methodological rules—in this case, rules of adequate explanation—if these do not contribute to its realization. An explanatory method which appears satisfactory or rational from the point of view of one ideal must be judged unsatisfactory or even irrational from the point of view of the other. If what counts from one perspective is a psychological understanding of the world of which we are a part, for the other it is the widening of our human ability to master this social and natural world technologically. It is a fact that contemporary science is a product of the institutionalization and realization of this second ideal. This does not mean that this ideal is eternal or the only possible one, or that it cannot be subject to critique. Such a critique will, however, prove unacceptable to those who believe that just as a knowledge of nature must serve as the basis for any effective technological domination over nature and allow us to manipulate objects in a controlled and predictable manner, so the knowledge of men and society should serve analogous engineering purposes.

This means, however, that the dispute about explanatory methods, and especially about ways of explaining social phenomena, is a consequence of the functioning of distinct ideals regarding what knowledge (in this case social knowledge) should be, and that no methodological arguments are able to convince an adherent of, for example, *Verstehenssoziologie* to accept the criteria of adequate explanation which an adherent of sociology understood, for example, as social engineering or social physics would be happy to accept—and *vice versa*. In order to accept the other's arguments, each would have to give up an ideal he has accepted.[19] Thus a necessary condition for the resolvability of methodological disputes is a certain *consensus omnium* which constitutes the shared ideal of scientific knowledge.

Generally speaking, it seems obvious that on the basis of an ideal of science which links inseparably the cognitive and the technological functions of knowledge, only those methodological rules which allow for a technological use of knowledge can be accepted and considered rational. I will discuss this problem in the following chapters. These chapters will also demonstrate that a general questioning of this ideal leads to a situation in which the nomological model of explanation has become controversial not only in the social but also in the natural sciences.[20] In addition to addressing the question of what constitutes rational methods of explanation within the framework of this ideal, we will also raise the question of the extent to which acceptance of this ideal can be considered rational at all.

CHAPTER III

THE MODERN IDEAL OF SCIENCE

1.

The development of modern science from about 1700 to the 1870s was the realization of a specific ideal of science which was different from the ideal of ancient or mediæval science. The new science, as Bacon had said, was to be *scientia activa et operativa*, a science whose goal it would be to equip human life with new inventions and means, supplying man with such knowledge of nature as would allow him to control it and use its forces. "We may find a practical philosophy" said Descartes, taking an epistemological position directly opposed to that of Bacon, "by means of which, knowing the force and the action of fire, water, air, the stars, heavens and all other bodies that environ us, as distinctly as we know the different crafts of our artisans, we can in the same way employ them in all those uses to which they are adapted, and thus render ourselves the masters and possessors of nature."[1] Thus it was supposed to be a science in which the cognitive function of knowledge was to be coupled with its technological function (which does not mean that the former was to be subordinated to the latter). Such a science would overcome the ancient opposition of *episteme* and *techne*, incorporating both at the same time.

We cannot ascribe necessity but only possibility to the existence of any object or the occurrence of any historical event. This means that no explanation can avoid referring eventually to some factual state of affairs, to something that actually happened, but about which we cannot say that it had to happen. No explanation can go beyond indicating "conditions of possibility." This means that on the basis of the same nomological structure, science always allows for a multiplicity of empirical worlds, and that the world which actually exists is the realization of one of a series of possibilities. Regardless of whether it is nomological or probabilistic, the nomological structure of the world which we discover and to which we appeal in explanation excludes the realization of certain empirical worlds which are nomologically impossible; but by itself it does not specify which of the nomologically possible worlds is actually being realized. This choice depends on so-called "boundary conditions," which have a random character from the point of view of a given nomological structure, which means only that they are not determined by that structure. Every specific explanation of events or processes must refer to these conditions which define the actual state of affairs. The elimination of this "duality of laws and boundary conditions" would be possible only if we were able to look at the universe as one whole without dividing it into distinct configurations: if we treated it as a system whose boundary conditions are just as necessary as the laws governing the system.[2] But this is something we are unable to do, whether we treat ourselves as parts of the system or believe that we can behave as external observers.

Accepting this view and inquiring how the modern ideal of science was born, I need not assume that its advent was necessary; I am asking rather how it became possible, and what circumstances can account for the fact that this particular possibility came to be realized. "There does not seem to be any sign that the ancient world, before its heritage had been dispersed, was moving towards anything like the scientific revolution, or that the Byzantine Empire, in spite of the continuity of its classical tradition, would ever have taken hold of ancient thought and so remoulded it by a great transforming power."[3]

Ancient science could, but did not have to give rise to its modern mutation. This is why we have to distinguish the question about those changes which were necessary for the formulation of the new ideal of science from the question of the conditions under which this ideal not only appeared but was accepted, institutionalized, and allowed to direct the intellectual efforts of scientists living within the sphere of European culture, which it molded in turn by imposing on it, among other things, its own models of rationality. It is a fact that this happened; but the claim that "it had to happen" is justified by nothing more than the cognitively barren principle that everything which actually happened was therefore necessary.

2.

Many historians and philosophers seeking to explain the emergence of modern science have claimed that it was born by abandoning philosophical speculation and by acknowledging the authority of experience and of the empirical world. I believe for a number of reasons that this view cannot be defended.

First, it is unjustifiable because of the empiricist character of Aristotelian philosophy, which for centuries stood almost alone in defining the direction and methods of the cognitive efforts of mediæval thinkers.

Secondly, we know how much effort and inventiveness the ancient and mediæval astronomers, for example, put into the task of matching their theoretical conceptions with their observations of the sky.

Thirdly, and most importantly, this view cannot be maintained because of the character of modern science, which did not arise from the generalization of empirical facts, as Bacon had wanted it to do, nor did it constitute an extension of the technological abilities of the middle ages and antiquity, as many historians have suggested when they treated Bacon's program for the renewal of the sciences as the actual story of its development, and as many sociologists have argued when deriving the emergence and development of modern science directly from the needs of practice. *Nota bene*: if the new science had really progressed according to Bacon's program, theory would never penetrate practice and could not have transformed this practice so as to allow for the emergence of applied sciences. At most it might have led to the use of theory in practice.

It is an indubitable historical fact that the process of the formation and development of modern science took place under the banner of observation and experiment which were to be separated from metaphysical speculations; that radical empiricism was an article of faith for many modern scientists, and that it was a program supported by numerous philosophers. The problem is that these views and programs do not correspond to the actual course of events, and that the "period is characterized by a kind of schizophrenia. What is propagated and declared to be the basis of all science is a radical empiricism. What is *done* is something different."[4]

Newton has often been taken as a representative and advocate of the empirical method. There is no doubt today that the famous saying *hypothesis non fingo*, with which he supplied his lack of an answer to the question of the nature of gravity, was not an account of the method by which he had arrived at the formulation of the principles of the new physics. By introducing such strange concepts as "momentary action at a distance," by differentiating absolute time and absolute space from relative time and space which was only apparent, and by endowing the whole of matter with a characteristic such as gravitation, Newton was not so much reporting on the experiments he had performed as building a conceptual apparatus with the aid of which physical experiments were henceforth to be interpreted. The famous polemics of Newton, first with the Cartesians and then with Leibniz, were not only disagreements between an empirical physicist and metaphysicians, but also a discussion among philosophers of nature who were arguing about God and the world on the ground of physics. If the metaphysician Descartes derived the law of momentum for physics from the idea of the immutability of God, then the empiricist

Newton derived the idea of absolute time and absolute space from the idea of the omnipresence of God in space, treating it as His sensorium. Moreover, the content of Newton's mechanics not only went beyond empirical data, and thus could not be derived inductively from this data, but was even at odds with some observational data known at the time it was formulated. There was for example no experimental criterion allowing for the distinction of absolute from relative motion. The case was analogical for other theories, including Galileo's physics, to which we shall return later.

What actually changed was the very concept of experience, with the aid of which scientists were attempting to link their theoretical constructions with facts; and it was this change which allowed for the emergence of a new ideal of scientific knowledge.

I do not intend to trace this process in the specific disciplines to show how, after beginning in physics, it eventually spread to all science. What seems most important and needs greatest emphasis is the fact that the emergence of modern science did not result exclusively or even primarily from the critique of old theories in order to make them more exact and consistent with the facts. What was at stake was not—either literally or metaphorically—an addition of new epicycles to the theoretical explanations put into question by observations. First it was necessary to "destroy one world and replace it with another. Scientists had to reform the structure of the mind itself, create new concepts and revise old ones, see existence in totally new categories, work out a new concept of knowledge and a new concept of science, and even replace the natural common-sense point of view with a different one which by no means appeared to be natural."[5]

Only on this basis, that is, as a result of the dissolution of the Aristotelian vision of the world and of man sanctioned in the middle ages, could the new ideal of science become possible.

In order to realize what this involved, we have to note the following aspects of the process of the formation of this ideal.

First, phenomena were no longer to be explained by reference to everyday direct sensory experience, to what is visible, but rather by experiments, sometimes even thought experiments, which were used to question nature.[6] This would be impossible without a questioning of the credibility of common sense, for which everyday experience was always the final authority. The new science is indeed a *scientia activa*, but not exactly in the sense of Bacon, of whom William Gilbert used to complain that he "writes philosophy like the Lord Chancellor." For Bacon, the term had a utilitarian rather than an epistemological or methodological meaning. This "active science" was to ask nature questions dictated by human needs.

In reality, the questions were dictated by theories, and the active character of science consisted in the creation of a system of categories in which the world was to be articulated, possibly so that objects could be manipulated. Thus, according to Bacon, problem-solving was to involve the registration, classification, and ordering of empirical facts, which constituted an extension of everyday practice rather than the theoretical questioning of this practice, the infusion of theory into practice. It is not necessary to prove that these two processes are not identical.

Secondly, mathematics and geometry were to provide the language in which scientific questions were formulated and in which the answers were to be interpreted. This was a result of the assumption, impossible on Aristotelian grounds, that the book of nature is written in the language of straight lines, circles and triangles; i.e., that it is a result of an ontological geometrization of the world and of the elimination of the gap between the heavens and the sublunar world. The choice of this new language in which science could ask questions of nature obviously could not be imposed by experience, since the use of this language, as a new conceptual articulation of the world, was precisely what was required for the new type of experience (based on experiment and measurement instead of what is directly visible).

Thirdly, the mathematical and experimental method of investigation which resulted from these changes required that the instruments used in experiments (and especially measuring instruments) not be regarded as extensions of the actions of our senses and hands in the way that tools are, but allow us to reach beyond what is directly accessible to the senses, and to interpret nature in quantitative categories. By the same token, instruments made it possible to study empirically objects which until then could not be subject to investigation, and—in the case of measuring instruments —to interpret the answers in the language of mathematics or geometry.

The use of an instrument causes a "splitting of the world" into those objects which are directly observable and objects with which we come into contact only through their representations as indicated by instruments. This leads at the same time to a "splitting" of cognition into everyday cognition, which deals with what is directly observable, and the cognition of those objects which are only indirectly accessible. Thus, Krzysztof Pomian distinguishes between visible and observable objects.[7] I would argue that the very distinction between these two categories of objects and two kinds of cognition is a result of the change in the concept of experience as it functions in science. Prior to this change, "observable objects" simply did not exist epistemologically as objects of experience. Although I would not want to identify modern science exclusively with the cognition that takes place through instruments, still there is no doubt that cognition mediated by instrumentation receives its legitimation and status only in modern science, and gradually encompasses broader and broader realms of reality: after physics, it entered chemistry, biology, and finally even the social sciences. The latter are to this day cultivated as the knowledge of both the visible and the observable, and this very fact lies at the source of many methodological controversies, such as those we discussed in the previous chapter. Moreover, the legitimation of cognition mediated by instrumentation appears to change our attitude towards direct cognition: we interpret the functioning of our senses in terms of instruments, and evaluate the data they supply in the light of various theories explaining their action.

Two philosophical problems which did not appear at all in the framework of the earlier worldview emerge as a result of this change in the vision of the world we have just discussed. The first issue is that of the actual existence or the ontological status of observable objects; and the second is that of the relation between direct and indirect cognition, or—in the terminology of Pomian—between everyday and scientific cognition.

As it happens, everyday language developed to meet the needs of everyday direct cognition is not very useful for the description of objects dealt with by indirect cognition. This concerns above all the qualities of size, shape, and relative position, or any characteristics which we determine by precise measurement. The emergence of a language to answer these needs, particularly the abstract language of mathematics, "goes hand in hand with the construction and use of instruments."[8]

This fact has made possible the gradual formation of completely new relations between theoretical and technological knowledge, and—as a result—the coupling of the cognitive and the technological functions of knowledge not only in philosophical programs but also in research practice. Without it, such conjunction would be impossible.

It was not the theoretical generalization of practical abilities, however amazing, that revolutionized science, but the design of precise measuring apparatus which introduced new methods to technology and made it— gradually of course—into applied science. The application of measuring instruments was a technical realization of theory rather than a tool gradually improved by trial and error; it led to the formation of a new language of science, to the separation of scientific knowledge from common knowledge, and to the current state of affairs in which "measuring instruments often are as large as factories, while factories work with the precision of measuring instruments."[9]

3.

The dissolution of the Aristotelian vision of the cosmos meant replacing the ideal of a finite, hierarchically ordered and ontologically diversified world, in which each being had its natural place where it properly belonged, with the idea of an open or even infinite world whose unity does not depend on the hierarchy of being, but on the sameness (universal validity) of the laws to which it is subject. Nature would now speak with one voice to everybody. The universe would be composed of elements, such as the earth and the heavens, all of which would have the same ontological status. In this universe physics and astronomy could no longer remain independent of one another; both would have to express the same universal mathematical or geometric order.

This eliminated from science all investigations based on values, perfection, harmony, sense or destiny. In the new ontology there is no place for all these anthropomorphic concepts, without which alchemy (for example) would have been impossible, but on the basis of which modern chemistry could not have been born. (And so alchemy ceased to be considered a science.) All such concepts were gradually perceived as subjective and eliminated one by one from the various branches of science: first from physics, then from chemistry, biology, and psychology. By the same token, references to final or formal causes ceased to be considered justified in scientific explanation, leaving only material and efficient causes as possible explanations. From the world of objects given to the senses, science was moving into a world in which abstract bodies move in an abstract geometric space according to universal laws. The world of sensual qualities accessible to direct cognition was being replaced by a world of sizes, shapes and relations, the world accessible to measurement. Direct cognition, which for centuries had been regarded as the authoritative measure of truth, was now considered misleading. In order to accept that the earth moves around the sun despite the testimony of our natural sense experience, science first had to eschew the world of common sense, for which sensual data constitute the final criterion, and to change this world in its own fashion, deciding that reality is not necessarily just what appears to our senses. What until then had constituted the basis of all explanation of phenomena had now itself become problematic and itself demanded explanation. Science was to express both this new world and the cognitive attitude of men to this world in new categories, changing the idea of human rationality and its criteria.

In this manner the Aristotelian idea that scientific explanation consists in reducing what is unknown to what is known came to be replaced with the conviction that the aim of science is to explain what is known by what is unknown; that is, the reduction of what is given directly to what is only "observable" and constitutes the hidden architecture of the investigated world. The more deeply the explanation reaches into a hidden reality, the more scientific it is. "Thus, scientific explanation, whenever it is a discovery, will be *the explanation of the known by the unknown*."[10] In the framework of the new ideal of science, the "depth" of an explanation itself becomes a cognitive value, and the striving for its realization constitutes the basis of the reductionist program of modern science.

The adherents of the view that the development of science is continuous often use very thorough and wide-ranging studies in order to argue that what I treat here as a revolution was the result of a long evolutionary process of introducing quantitative methods into various areas, which began as early as the twelfth century. Without questioning the facts discovered by these valuable studies of the science and culture of the Middle Ages (studies thanks to which we can without hesitation reject the view that the Middle Ages was a period of total suspension of human inventiveness and curiosity), we can nevertheless say that the facts cited tend to undermine rather than support the thesis they are supposed to defend.[11] Despite the undeniable genius of people such as Grosseteste, Buridan, Nicholas de Cuxa and Nicholas d'Oresme, and later Benedetti or Tartaglia, they were unable to solve the problems they attacked because these problems could not be formulated properly on the basis of the Aristotelian ontology, which none of them challenged. It is enough to read, in *De Revolutionibus*, the answers of Copernicus to the physical problems supposedly resulting from the idea of a revolving earth, in order to realize that all of them were directed against him and that he could not refute them because he had to defend himself on the grounds of his opponents, that is, in terms of Aristotelian ontology and physics, and on these grounds it was impossible to defend the Copernican theory. As the history of the discovery of the law of free fall or the law of inertia demonstrates, Galileo was also unable to solve these problems until he rejected this ontology.[12] The heart of the matter is after all not the question of whether attempts to use quantitative methods in physical investigations were made before Descartes and Galileo —since they clearly were—but rather that the conceptual articulation of the world did not allow scholars to use this method effectively, so that their attempts to use it resulted in failure. A fundamental change of basic concepts, above all the concepts of time, space, and experience, was a precondition and not a consequence of the success of such attempts. This is what I have in mind when I claim that what took place was a revolution.

The fact that it took place over a period of time rather than instantaneously goes of course without saying.

The situation was apparently similar in disciplines other than physics, and even outside the natural sciences.[13] The discovery of oxygen gave rise to a new chemistry; but Priestley's discovery did not and could not have any consequences for chemistry so long as the process of combustion was treated as an example of the decomposition of a substance rather than as a process of synthesis. In order to cease regarding it as decomposition and to develop the idea of using measurement (weighing the substance before and after combustion), it was necessary to question the credibility of sensory experience—the escape of smoke and fire during burning. It was necessary to accept that the world is not exactly just the way it appears to our senses, and that its hidden architecture can be subject to number and measurement. What prevented the alchemist from becoming a chemist was not exclusively and not even primarily a lack of measuring apparatus, of a scale or a thermometer (the scale was perfectly well known and used for example by goldsmiths). What was missing was above all a view of the world which would sanction the use of such instruments. Once the alchemist had adopted such a view he became a chemist, ready to use the scales, the thermometer, and hundreds of other instruments which he designed himself on the basis of accepted theories, and by means of which he would organize his experiments. This is why I accept Koyré's claim that the destruction of the Aristotelian vision of the world, i.e., *the reform of natural philosophy*, was a precondition of the formulation and use of the mathematic-experimental method of investigation. This is also why the emergence of modern science cannot be treated as the result of a gradual growth of knowledge within the framework of the old ideal of cognition. Koyré's studies provide strong historical evidence and theoretical justification for this view, which explains, among other things, why people who were justly considered the best minds of their time subscribed to views which today can be refuted by any high-school student on the basis of familiar empirical facts which were also known then, but which were then interpreted in terms of completely different ontological categories. This also explains why a high-school student or a layman finds it so difficult to understand or even believe that outstanding scientists could maintain for centuries such views as, for example, that an arrow shot from a bow is kept in motion by the pressure of the air whose motion was initiated by the release of the bowstring, when they knew perfectly well that an arrow could also travel against the wind. It also explains why the revolution discussed here was so difficult, complicated, and long lasting.

These bygone views appear nonsensical to us if we remain unaware of the extent to which our vision of the world, and the ideal of science with which it is associated, differ from those earlier visions which we now reject, often making little effort to understand their place in the image of the world current at the time. But all that this means is that our own vision of the world and its associated ideal of scientific knowledge appears to us already as something natural, obvious, and unproblematic.

4.

A new ontology was, however, not the only necessary condition for the emergence of the modern ideal of science. From the epistemological point of view, the revolution required the replacement of faith in the supreme value of tradition and sanctified authority by a conviction that man is capable of learning the truth by using his own native senses and the inborn capacities of his mind, and moreover, that progress in this sphere constitutes a positive value. In other words, what was necessary was the idea that a subject—at least insofar as he functions as a knowing subject— can be completely autonomous: independent of tradition, of personal experience, and of all specific circumstances. Such a free subject can find in himself, in his own cognitive capacities, the basis for constructing scientific theories which will be always and everywhere valid. His reason enables him to discover alone, through his own efforts, the universal character of the natural order. *The modern ideal of science would be impossible without the conviction that an individual and physically limited subject is able intellectually to overcome his particularity and achieve universally valid knowledge.*

Only on the basis of this conviction common to various modern philosophical directions could divergences among the methods guaranteeing the rationality of the subject be articulated. These divergences concerned the methods leading to knowledge, thanks to which the subject accepts only those results of his cognitive activities which must also be accepted by others, and thanks to which he will consider as true only that which always and everywhere must be considered true.

As an autonomous knowing subject, man has to be equipped by the philosophers with at least a few divine attributes. At least potentially, that is, independently of the various accidental circumstances which might deform his cognition and of various idols (whose influence could, however, be neutralized), a subject must be able to be an ideal observer *who is excluded from the world he is investigating, whose cognitive processes do*

*not depend on his position in this world, and whose investigative activities
and results in no way influence the object under investigation.* Moreover,
he must be able to engage in the most complex forms of logical reasoning
on the basis of the knowledge he possesses. The demon described in
Laplace's treatise on probability can serve as a model of such an ideal
knowing subject. In any case, beginning with Descartes and Bacon, this
notion of the cognitive autonomy of the subject was articulated in various
versions of the scientific method, which it was believed could be established and decreed once and for all, since it corresponded with the
immanent capacities of human nature. At the same time, this idea co-
determined the concept of experience and the concept of truth, and set the
possible limits to human cognition. The entire intellectual effort of the
scientists' domination of the world took place within the framework of this
conception.

In this context, I would like to raise two issues.

First, *the ideal of an autonomous knowing subject constitutes a
philosophical justification of the social postulate of the autonomy of
science as an institution*: of its independence from the church or the state,
and of the independence of investigation as a method for knowing truth
from all religious, philosophical, or political views accepted beyond the
realm of science. Such independence allowed the scientist credited with
the ability to discover objective truth to pretend to the role of an impartial
arbiter in all sorts of controversies which could be solved with the help of
the scientific method. This exceptional position was accorded the scientist
as an autonomous knowing subject by the use of a special method of investigation.

The empiricist philosophy of Bacon and his followers which, as we
have seen, did not provide the best account of the way in which the new
ideal of science was being realized, owed its success in large measure to
the fact that it sanctioned the autonomy of scientists in society, presenting
their activity as neutral and unbiased by any prior beliefs, with generalization from empirical facts constituting the unquestionable foundation of
knowledge. According to this philosophy, the scientist armed with an
appropriate scientific method faces nature directly, bringing nothing from
himself into the encounter, but instead drawing logical conclusions from
indubitably established facts which—once they are confirmed intersubjec-
tively—can never be altered: theories can change, but facts once discovered
remain facts forever; they are the invariants of our knowledge.

This aspect of the empiricist philosophy seems to explain why so
many scientists, including many distinguished ones who in their own
investigations did not and could not follow the principles of this method,

presented themselves as its adherents and claimed that their results were the outcome of applying this method. Newton's example is perhaps the most spectacular, but he was by no means the only one. In any case, the contributions of modern empiricism from Bacon to Carnap to the defense of the autonomy of science appear incomparably greater than its actual influence on the method of conducting scientific investigations. The empiricist philosophy was the basis of a particular ethics of cognition: it commanded humility towards facts.[14] Unfortunately, it also commanded humility in the face of what only passed as facts.

The epistemological conviction according to which it is possible to engage in cognitive activity in total independence from all external factors, and the related notion of the autonomy of science as a social institution, corresponded for a long time to the actual historical situation of science in society. Generally speaking, this situation was characterized by a lack of institutional connections between science, economics, and politics. When this situation began to be treated as universally valid, reflecting the very nature of scientific activity, it promoted the idea that the modern ideal of science should be treated not as a cultural fact but as a natural fact, which found its expression in the ideology of scientism. We will return to this when we discuss changes in this specific historical configuration and when we consider the problem of how the development of knowledge itself began to undermine the ideal of the autonomy of the knowing subject. Eventually this ideology revealed more and more false consciousness and began to disintegrate.

The second comment concerns the fact that *the ideal of the autonomy of the knowing subject reduced the problem of human rationality to a purely epistemological issue.* Regardless of whether it was argued that human rationality should be based on the acceptance of only those claims which a person is able to verify with the help of his own innate cognitive powers without appealing to faith or authority, or of only those statements which are subject to the immediate control of experience, or of those assertions whose necessary character can be demonstrated through pure reasoning—all of these conceptions of human rationality were underwritten by a model of human nature as (at least potentially) perfectly rational; that is, they rested on the assumption of the basic autonomy of cognitive activities rooted in human nature and considered rational. By the same token, the distinction between rational and irrational procedures—between procedures which are and are not worthy of confidence as methods for gaining true knowledge—was also to be treated as a purely epistemological problem. Were it not for the conviction that at least in some cognitive activities, the subject can be autonomous with respect to both the natural

and the cultural world, it would be impossible to formulate the criteria of
rationality in purely epistemological terms; and were it not for the idea of
the scientific methods which codified all these cognitive procedures, it
would not be possible to believe that the scientific knowledge which results
from their use is the embodiment of human rationality. It is thus not
surprising that whenever the idea of the cognitive autonomy of the subject
and of the autonomy of science as a social institution comes to be questioned, the rationality of science and its development becomes the focus of
attention in philosophical reflection on science.

Without anticipating our later argument too much, we can say here
that the philosophical revolution which made possible the articulation of
the modern ideal of science consisted in closing the chasm between the
earth and the heavens; it replaced the world of qualities accessible to the
senses in which we live, feel and think, with the world of quantities and
shapes; and gradually but inevitably it eliminated anthropomorphism and
anthropocentrism together with all ideas based on values, or on the immanent meaning of the world. It established a new conception of the
knowing subject who, thanks to his autonomous cognitive powers, was supposed to be able gradually to learn the truth which would be always and
everywhere valid; and the conception of science came to be understood as
precisely this form of knowledge.

At the same time, the development of knowledge taking place for
over three hundred years within the framework of this new ideal of science
was more and more clearly constructing another chasm: the chasm between
the world of facts and the world of values, which science could neither
explain in its own categories nor dismiss. Practical reason may and indeed
does attempt to bridge this chasm, but it remains unbridgeable for theoretical reason. The Kantian dictum, "The starry heavens above me, the moral
law within me," was a dramatic statement of this split, which the
Newtonian synthesis made fully visible and which is still with us to this
day.

5.

For many centuries, scientific and technological thought developed
independently of each other: scientists were not terribly interested in
technological problems since they were convinced that they could not be
solved in a scientific manner, while artisans and engineers had no special
need to be interested in theory.

This separation of theory from practice is often explained with reference to the social structure, with its aristocratic disdain for physical labor and earthly matters. The source of modern science is then identified as the change in this sphere which took place in Renaissance Europe as the result of greater interest in technology, practical life, and the worldly dimension of human existence. These explanations surely contain some truth; I think, however, that they have not much to do with the articulation of the new ideal but serve rather to indicate the necessary conditions for its social acceptance. They explain why the reformation of natural philosophy could become a historically important phenomenon, and why it did not remain merely the aberrant dream of a small group of prophets (how many reforms have ended this way?), a historical curiosity without greater significance. If the interest in practice alone were a sufficient condition for the formation of a new ideal of science, if the new science were really just an extension or theoretical generalization of the practical activities of artisans and inventors, then—as Koyré notes—it should have appeared much earlier as the work of the engineers of the Roman Empire. Just as the philosophically oriented thought of the Greeks with its ideal of epistemologically certain cognition (*episteme*) failed to give rise to applied sciences, so the more practically oriented culture of Rome also failed to generate a new ideal of science.

Although cannon balls destroyed mediæval castles and contributed to the fall of feudalism, mediæval mechanics did not follow as a result. The mechanics of Galileo was not born from a generalization of the experiences of artillery experts, shipbuilders, or arsenal keepers in Venice, although Galileo was indeed interested in their work. Nor was it created for the purpose of improving their abilities, though in the end it did contribute to such improvement. In any case, while Galileo might teach ballistics to artillery engineers, he was certainly not drawing scientific conclusions from their practical skills.

The scientific method created by Galileo did contain the seeds of the possibility of giving a new direction to technological thought, by subordinating it to the demands of a theoretical precision which until then was believed applicable only to *episteme*. The process of the realization of this possibility defined among other things the history of European culture between the first industrial revolution which took place largely without the involvement of science, and the scientific-technological revolution resulting from the direct utilization of scientific theories as the basis for technological activities. In short, it is necessary to distinguish the possibilities contained in the new method from the conditions of their realization.

Or, according to Koyré:

> What could be simpler than a telescope or a field glass? To build them one needed no theory or special lenses (unlike those used in eyeglasses), that is, no highly developed technology. It was enough to place two lenses from eyeglasses one on top of another, and a spy-glass was constructed. How amazing it seems that for four hundred years no one had ever been interested in the question of what happens when instead of using one pair of glasses we use two simultaneously.
>
> But glasses were not made by an *optician* but by an *artisan*. He was not constructing *optical instruments*, but *tools*, and he did so in accordance with the traditional rules of his profession, without striving for anything new. The legend which teaches that the invention of spy-glasses was a result of *chance* since they were invented by a child of Dutch artisans who was playing with lenses, contains a deep truth, even if the story is a legend.[15]

But while the Dutch artisans who had the idea of connecting lenses into spy-glasses were busy trying to bring technological improvements to the invention (linking the lenses in a tube, moving the eye-piece), Galileo, the moment he learned about the Dutch field-glasses, tried to construct its optical theory. On the basis of this certainly imperfect theory he tried to improve the precision and reach of the lenses, in order to build his telescope, and then a microscope. The Dutch spy-glasses were designed to bring closer what was in any case visible to the naked eye (*visible* objects), and for no other purpose. It is no accident that the Dutch designers and users did not even try to make their glasses useful for observing what is is not directly accessible to the eye, either because it is too distant or too small. Galileo, on the other hand, built his optical instruments and pointed his telescope into the sky in order to see what was not directly visible, what no one had ever yet seen, and in order to meet the theoretical needs of astronomy and physics. In order to build such instruments, it was necessary not only to solve the technical problems connected with the making of lenses of appropriate quality, but also to determine in advance the proper shape of the lenses and to build a machine which could give them this shape with the required degree of precision. In this respect, it is no accident that the first optical instrument was invented by Galileo, while Descartes designed the first machine to polish parabolic lenses. These were perhaps the first machines whose functions were to be governed by the demands of theoretical precision.

The matter was similar with instruments for the measurement of time. Mechanical clocks, whose invention brings well-deserved credit to the technological thought of the Middle Ages, were initially not even as precise as the water-glass of the ancients. At the same time, they were enormously expensive and only big, rich cities could afford them. In the second half of the sixteenth century the situation changed: thanks to numerous improvements mechanical clocks became much more exact, began to be used more widely, and began to mark the rhythm of everyday life, especially in the growing and prospering cities. And yet, use of the clock as a measuring instrument cannot be derived from these improvements; it was not an extension of the inventions of artisans, but the realization of a scientific theory.

In order to measure the passage of time with some exactitude, it is necessary to appeal to a process which is completely monotonous or repeats cyclically. The first possibility found expression in the water glass, from which water flows out at a constant rate thanks to the maintenance of a column of water of constant height in the tube. The second solution, adopted by Galileo and Huygens, exploits the isochronous movement of the pendulum. It is obvious that a discovery of this kind could not be the work of sheer empiricism, since in order to establish whether water indeed flows out at a constant rate or whether the pendulum is isochronous, we need a measuring apparatus.

When Galileo conducted his famous experiment with the inclined plane, he had to measure time with the help of a water glass. It is not surprising that the results were inexact (*nota bene*, every radical empiricist should ask himself here how Galileo was able to formulate his laws at all without the help of an exact clock!). Koyré notes pertinently that Galileo must have been concerned with the problem: "why do we need mathematical formulæ allowing us to determine the velocity of the fall at every moment in relation to the acceleration and the time of the fall, if we are unable to measure either the one or the other?"[16] Despite the legend, Galileo was unable to ascertain the isochronous movement of a pendulum by observing the candelabra swaying in the cathedral of Pisa, nor could he formulate the law of free fall by dropping weights from the celebrated leaning tower. Both legends are the products of historians' imaginations, originating in an extreme empiricist conception of the development of science and the explanation of its history on this basis.[17] Incidentally, the candelabra in Pisa cathedral appeared only after Galileo had left the city, and the weights were dropped from the tower by one of Galileo's opponents. The results, obviously, argued against Galileo, which he countered by noting the lack of precision in the measurement of the time of the fall.

Characterizing Galileo's method, Torricelli wrote: "*I imagine and suppose* that some body is moving downward and upward according to a known proportion, and horizontally with uniform motion. [...] If then balls of lead, iron, or stone do not observe that supposed direction . . . *we will say that we are not speaking of them.*"[18]

This means that by using this method we move from the world of visible objects existing in physical space and subject to directly observable motion, to the world of abstract constructs—for example, material points —moving in a geometric space in which motion can be analyzed into its component parts. This is no longer the world of objects which can be perceived with our senses, but of objects which can be measured, whose real existence is assumed by theory and tested in experimental measurement. Indeed, Galileo and his followers speak of something other than lead, iron and stone balls which can be felt with the senses. *But a philosophical revolution was necessary in order for this "speaking of something else" to become legitimate in science,* in order for experiment to be defined by theory, and for the results of experiments to be criticized and interpreted in theoretical categories rather than in the categories of everyday experience. For anyone who did not accept this revolution, who remained in the world of the Aristotelians, the statement "too bad for the facts" must have sounded horrendous. And let us add: the conviction that not the objects immediately given to sensory experience, but abstract constructs were to be the objects of scientific theories did not become accepted in all disciplines at the same time, and almost always met with fundamental opposition, which in some areas continues to this day.

Let us go back, however, to the isochronism of the pendulum. Galileo deduced it from his theory of accelerated motion for the case of a weight tied on a string, that is, moving in a circle. In the same manner, Huygens determined later, as against Galileo, that isochronism does not occur in motion in a circle but rather in cycloidal motion.

Only on this basis, with reference to a theory indicating which movements of real bodies could approximate isochronism, could the technical problem of the realization of a mathematical model arise, namely the problem of building a precise clock: a chronometer suitable for scientific measurements. In order to achieve this it was necessary to "teach 'technicians' to do something they had never done before: to impose new rules on craft, on art, or on *techne*, rules of precision of *episteme*."[19]

Instead of beginning with what is immediately given and abstracting qualitative regularities, cognition should begin from mathematical theories and deduce from them hypotheses directing the action upon objects. This is both the road by which a mediæval artisan was transformed into a modern engineer, and the road to an *active science*, a science which could not only be used in practice, but could fundamentally transform it.

Similar changes were occurring at that time in other areas of craftsmanship and practical skill as well. Painters learned the rules of perspective from mathematicians: "the force of lines and angles," as Piero della Francesca used to say. Geographers and cartographers learned from them the rules of triangulation; architects learned statics. Discussing the building of the dome of Santa Maria del Fiore by Brunelleschi, Pierre Francastel writes:

> It was no longer a matter of making calculations on the grounds that the profile of one stone is decided by the profile of the neighboring stone, and in extremity it is possible to try the fit out on the scaffolding. Now the problem was to determine with the help of abstract calculations the angle and location of numerous small elements such as bricks, taking into account their double function, as the carrying skeleton and as filling material, without any possibility of empirical control. The work of Brunelleschi defines the moment of transition from the phase of empirical technology to the stage of mathematical speculation: the Renaissance builder becomes an intellectual, while the builder of the middle ages was an artisan.[20]

This is what I meant by suggesting that it was *the measuring apparatus whose application was demanded by the new ideal of science which became the link between scientific and technological thought* and made it possible to transform a technology to which such rules of precision had always been foreign into a technology based on these rules. The first precise machines were made for the production of measuring instruments; and the construction of precision instruments became the first industry in the middle of the sixteenth century.

All of this together points to the most important conditions which had to be met before the new ideal of a science linking the technological and the cognitive functions of knowledge could become possible.

6.

It is impossible here to discuss the various social processes which took place in seventeenth-century Europe and which, by altering its culture, rendered the old ideal of scientific knowledge incapable of realizing the values of the new culture; while the new ideal resulting from the postulated

revolution in natural philosophy could become socially accepted, institutionalized, and implemented in various areas of research, causing what
today we traditionally refer to as the scientific revolution. I would only
say that any attempt to explain this fact with reference to a single *principal
cause* seems to me to be hopeless, regardless of whether this cause is
understood to be the development of cities, the rise of the middle classes,
the Reformation, developments in technology and artisanship, travels and
geographical discoveries, the formation of centralized nation-states, or the
secularization of culture. It seems impossible not to view all these listed
and unlisted changes together as various aspects of the global historical
process of cultural change, of a process which made the realization of the
new ideal of knowledge possible, the course of which was in turn becoming more and more dependent on this revolution. No phenomenon of
this sort has a single cause, and no effect is a mere epiphenomenon which
does not in turn influence the fate of the system which gave birth to it.

All historical and sociological conceptions which have assumed that
one such deciding factor exists and can be revealed, no matter how cleverly
they have described the role of the given factor, have always turned out to
be one-sided, although—obviously—a detailed analysis of the effects of
any such factor usually contributes to a better understanding of the global
process. As a rule it would appear that the links between intellectual work
and its social context cannot be explained by any single factor, or that this
factor decides only in the last instance, as Engels claimed.[21]

In view of my goals in this book, it is enough to say that I believe
that this global cultural process constituted a kind of filter allowing for a
"natural selection" of the ideal of scientific knowledge, a filter which
determined that certain ideals would die out because they ceased to be
functional in view of the changes taking place in the "inherited set of
values," while others gained in social acceptance and became historically
important for the same reason.

In this context it is worthwhile noting that the intellectual changes
taking place in the fifteenth and sixteenth centuries in Western Europe,
under the slogan of the rebirth of antique culture and of a return to the
sources of faith, by no means led to a resurrection of the antique ideal of
knowledge, but to the formation of something entirely different from it.

The seventeenth century debate between Ancients and Moderns did
not directly concern the development of the philosophy of nature and
science, but of literature and art. In essence, however, it was a debate
about the concept of progress central to all of modern culture. At issue
was the question of whether—as the Ancients claimed—the changes that
were taking place in culture should lead in the best of cases to a "return"

to the culture of antiquity, which appeared as the highest imaginable achievement of humankind, or whether—as their opponents claimed—they should lead to constant progress, in which ancient culture constitutes only one stage to be surpassed.

The adherents of the new science, even when they appealed to Plato and his followers in order to defeat the Aristotelians (and neo-Platonism influenced Copernicus and Kepler as well as Galileo), opted in an obvious fashion for the second of these alternatives. And indeed, the realization of the ideal of the new science was to constitute perhaps the strongest affirmation of this new conception. If, as Fontanelle wrote, in the art of expression and poetry, which depends on imagination, the Ancients could even surpass the Moderns, in science there is progress, so that "the last physicists and mathematicians are perforce the most able."[22]

Thus, regardless of whether the new ideal of science founded the idea of progress, or the acceptance of this idea helped to establish the new ideal of science (and the one does not exclude the other), it does not seem possible to answer the question of why a real renaissance of the ancient ideal of science did not take place; and the question why the call for its rebirth should have resulted in the formation of a new and different conception cannot be answered without going beyond the internal history of scientific ideas. The linkage between the cognitive and the technological functions of science was, after all, by no means an epistemological necessity.

"Prescientific cultures," according to Ossowski, "were presumed to be unchanging. What was valued most in them was that which could claim the status of complete stability. If—for example—religion did change, then this would happen only in such a manner that no one would be fully aware of it (by the gradual transformation of religious traditions), or through a revolution in which one "unchanging" religion was replaced by another pretending to a similar constancy. In contrast, the modern scientific culture is not only in a state of permanent change, but this dynamism is taken as a postulate by those who contribute to it."[23]

The idea of incessant progress underlying modern culture since the renaissance has meant that even the most disquieting and unexpected changes in the content of scientific knowledge have not constituted in this culture something foreign and impossible to tame. In this respect ours is an omnivorous culture. And what is more: the very possibility of such changes is written into its model. Only a questioning of the very idea of progress, especially technological progress, for which modern science has become the primary engine, could lead to a situation in which the modern ideal of scientific culture could become controversial.

In summary, I am inclined to say that *if the formation of the new ideal of science was made possible by a reformation of natural philosophy,* as we saw in this chapter, *then the conditions making possible the acceptance of this model involved changes in the inherited system of values passed on from generation to generation, in which the notion of progress, and particularly technological progress, played a central role.* These changes led to the birth of modern culture.

The drawing of a distinction between these two processes—of the formation and the acceptance (institutionalization) of the new ideal of science—is obviously a methodological procedure rather than an attempt to distinguish some "historical stages" of development. The distinction, however, does appear both useful and necessary. It allows us to see that the necessary conditions for the intellectual articulation of a new conception of cognition are different from the conditions of its social acceptance and realization, conditions without which the new ideal of science could never have transformed actual investigative practice to the extent of becoming in its turn something "obvious," "uniquely rational," or "almost natural," as the modern ideal would soon begin to be seen.

CHAPTER IV

THE INSTITUTIONALIZATION AND PROFESSIONALIZATION OF SCIENTIFIC RESEARCH

1.

The modern ideal of science, and the mathematical and experimental method of investigation which contained the seeds of the subordination of technology to the demands of precision, did not lead immediately to a direct linkage between science and the economy. In any event, the industrial revolution of the eighteenth century took place practically without the participation of science. It was brought about by inventors, people of a practical bent who were not always and not necessarily acquainted with the theoretical basis of their inventions. These inventors worked usually by the method of trial and error. They proceeded empirically in the bad as well as the good sense of the word, and often despised theoreticians, regarding them as divorced from real life. Theoreticians paid them back in kind, claiming that they were not disinterested and acted not with the aim of advancing knowledge, but for personal and material gain, which is improper for men of science. This period can be virtually symbolized by the names of James Watt and Thomas Edison. The steam engine, which literally and metaphorically served as the engine of the industrial revolution, was constructed well before Fourier, Carnot, Clausius, Maxwell and Boltzmann established the foundations of thermodynamics. Their theories later served to introduce various improvements in the invention, but the innovation itself not only appeared but also found widespread use without their help. Lewis Mumford claims that "The detailed history of the steam engine, the railroad, the textile mill, the iron ship, could be written without more than a passing reference to the scientific work of the period."[1] This judgment certainly does not apply to the history of technological development during the last hundred years. Edison's greatest invention, as Norbert Wiener notes ironically, was "the industrial research laboratory, turning out inventions as a business";[2] but priority in this field probably rests with the chemical industry of Germany and England.

The "delay" in the realization of the possibilities offered by the modern ideal of science resulted from the fact that the conditions necessary for its realization were met only towards the end of the nineteenth century. It is certainly a historical irony that it was precisely this achievement which

65

in effect led to the crisis of the modern ideal of science in the twentieth century. This crisis has not slowed the pace of intellectual development—which, on the contrary, is faster today than ever before in history—nor has it limited or weakened the role of science as a social institution. But the conception of science formed in the preceding period, which had an overwhelming appeal in the community of scientists and in society at large, began to conflict more and more obviously with reality. It did so both because of changes in the place and role of science in the social structure, and because the cognitive achievements of science appeared to undermine the epistemologically grounded assumptions on the basis of which science had been developing for three hundred years. Today this conflict finds its expression both in the scientist's self-knowledge and in the philosophical reflection on science.

Generally speaking, the crisis of the modern ideal of science has two sources. First, *it results from the undermining of the autonomy of science as a social institution*, which leads to deep transformations in its internal structure and ethos, and in social trust and confidence in science. Secondly, *it results from the undermining of the concept of the cognitive autonomy of the subject*, allowing for the achievement of totally objective knowledge, valid always and everywhere, and independent of the location of the subject in the cultural world. As a resuslt, questions have been raised about the exceptional character of science as a special kind of knowledge and a special method of achieving truth. Let us examine both of these issues in greater detail.

2.

From a sociological point of view it could be argued that science as we know it today was shaped by two processes. The first can be defined as a process of institutionalization; the second as a process of the professionalization of research linked to industrialization. This second process produced a fundamental change in the role and place of science in social life, and brought about changes in its internal institutional structure.

Like Joseph Ben-David, I assume that an activity is institutionalized and becomes a social institution or a social system under the following conditions.

First, the society or some part thereof must accept this activity as fulfilling an important social function considered valuable in itself.

In the case of science this meant the acceptance of research as the legitimate method of gaining knowledge which—like art—contributes to human self-understanding and to an understanding of the environment, while being essentially different from tradition, philosophical speculation or revelation. At the same time, it implied that those engaged in scientific research were recognized as having the authority to determine the characteristics of this means of gaining knowledge and to identify its trustworthy achievements. Acceptance of the knowledge gained through scientific research as universally valid and valuable was based, as we have seen, on the idea of the cognitive autonomy of the subject.

Secondly, the institutionalization of any activity requires the formulation of norms regulating the conduct of those engaged in it. Such norms are supposed to guarantee the realization of the goals of this activity and its autonomy vis-à-vis other social systems.

In the case of scientific research this meant that scientists had to subscribe to such norms as:

a) a disinterested search for truth;

b) making the results of their work known publicly, both in order to make possible control and critique by others and in order to make these results available for use in further investigations or in the development of possible practical applications. Scientific knowledge was not to be secret, licensed or patented; it was to be considered common property and a common good;

c) acceptance of the claim that the value of scientific statements does not depend on the identity of their author. Respect for this norm made possible the formation of an international community of scientists in which the position of each scientist would depend only on his achievements as acknowledged by his peers; and

d) skepticism towards the results achieved by others; that is, accepting personal responsibility for using results obtained by others and making public all criticisms and objections to other people's results.

Acceptance of these norms constitutes the condition for the efficient functioning of the system regardless of whether or not one believes that its goal is—as Merton claims—the maximally rapid growth of new knowledge, or—as Storer believes—the achievement of public recognition, which functions in the system as a specific resource in exchange for personal achievements. In either case, these norms have been seen as determining the internal conditions of cooperation and competition among members of the system. They have also constituted the basis for the construction of the ideal image of a scientist as someone who, at least in his or her intellectual activity, is guided exclusively by these norms. The extensive literature popularizing science has propagated precisely this image of the scientist.

Thirdly, institutionalization demands some adaptation of the norms regulating the behavior of scientists to the functioning of other social systems and to the norms by which other activities are governed.

In the case of scientific research, this means that the social system must tolerate conditions such as freedom of speech and freedom of opinion, freedom of communication, and at least some tolerance of religious, ideological and political differences, as well as allowing for social changes which might follow from free investigation and its applications. The efficient functioning of science as a social institution demanded not only that the norms named above had to be accepted in the scientific community, but also required an internal arrangement in which the values to be realized by following these norms would not come into direct conflict with values which the more global social structure is expected to realize. If these values could not be accepted and defended effectively by people other than scientists, the scientific community would find itself in a state of permanent conflict with the rest of the society and could not exist there, at least not for long. It would either have to disappear or, what is more likely given our contemporary experience, it would have to undergo essential transformations.

Many contemporary manifestations of the pathology of scientific life result from such conflicts between values whose realization constitutes a condition necessary for the functioning of the intellectually autonomous scientific community, and values being realized either by the society as a whole (in totalitarian systems) or by certain other social institutions on which modern science depends everywhere, even in pluralistic systems.

The optimal condition for the development of science would probably be a situation in which both systems of values were identical, and many authors have believed that the development of science might eventually lead to such a situation, given the value of science as a factor central to the process of the rationalization of social life and social conflict. It might be superfluous to add that contrary to such scientistic utopias, such conditions have never existed anywhere, and it is difficult to expect them to do so in the future. The vision of a Brave New World today seems far more realistic than the vision of a new Atlantis.

In any case, thus understood, the process of the institutionalization of science began in Western Europe by the sixteenth and seventeenth centuries and later spread elsewhere. The later the period in which this process took place in a given country, the more it was linked to the simultaneous process of professionalization, and the weaker became the native tradition of an autonomous science independent of other social systems.

Everywhere, however, the process of the institutionalization of scientific research was accompanied by the formation, around the group of scientists in the strict sense of the word, of a milieu of educated laymen interested in science and able to follow its achievements and to participate in scientific controversies. The institutionalization of science "could not take place were it not for a crowd of erudites who picked up controversial problems, commented on the arguments of both sides, gave them philosophical interpretations, and welcomed with enthusiasm the rebirth of the scientific spirit."[3] It was chiefly this intellectual milieu which at the time determined the social acceptance of science as an important element of the new culture. The first scientific societies and academies, which in the early periods of their existence united not only practicing researchers but also people interested in the results of scientific investigations, were established in part thanks to this group of intellectuals. As late as 1840, there were only 100 scientists among the 600 members of the Royal Academy, while the rest were gentlemen interested in science, army and navy officers, pastors, lawyers and doctors. Even in the council of the society, scientists achieved a majority only in the 1820s, during Davy's presidency.

The formation of academies and scientific societies was accompanied by a gradual transformation of the private correspondence of scientists into a system of publication in the scientific periodicals which these academies and societies began to publish. (*Philosophical Transactions* began appearing in 1665, the *Journal de Savants* in 1666.) This system guaranteed the popularization of achievements, the possibility of broader control, and the formation of an international scientific community. Especially in the natural sciences, the publication of articles in journals gradually became the primary medium of communication of new results, which until then had been announced through scientific treatises.

The process of institutionalization also led to the reform of the universities, whose structure had been established in the Middle Ages and which remained to a large extent dependent on the Church. These reforms —which took place in England in the seventeenth century, in France during the Revolution, and under Napoleon and in Germany after 1809 (that is, after the establishment of the University of Berlin), obviously took place under various political and cultural conditions which I cannot discuss here.

Common to all these changes was the fact that scientific research, as opposed to teaching, was still only a private concern of individuals rather than a responsibility derived from the occupation of a particular position in the existing organizational structures. There were simply no organizations whose main task was to conduct scientific research. Neither the scientific societies nor the academies and universities constituted such organizations, even if they occasionally supported individual scientists. In this sense the scientific investigator remained an amateur. Scientific research, like art, was meant to be, and as a rule was, a *spiritual calling rather than an occupation*. At best the scientist, like the artist, could rely on private patronage. The occupation of researcher is a product of the twentieth century, and its emergence is the result of a basic change in the place of science in social life.

An amateur scientist could be a member of the wealthy classes and treat his research as a hobby: this is why the careers of such people as Davy or Faraday were exceptional and spectacular. A scientist could, as in the case of Lavoisier, occupy a position in state service which did not demand that he conduct research, but which rendered his research possible, and he could engage in research on his own account. He could also, as was often the case, and in Germany was almost the rule, become a university professor and earn his living by teaching. A university position offered some possibilities for conducting research; chairs were usually granted only on the basis of earlier scientific achievements, but they did not oblige the teacher to continue such research.

This amateur status of the researcher corresponded with a situation in which science, apart from a few sporadic contacts, was not linked with economics or politics by any formal relation. The need for new theoretical knowledge beyond the scientific community itself and a narrow circle surrounding it was basically non-existent, and their fellow scientists were—at least in the minds of the researchers themselves—the only audiences for scientific work. The solutions to problems proposed by one scientist were to serve as a starting-point for the research of others, and new knowledge was prized primarily as an autotelic value. Throughout the eighteenth and nineteenth centuries, the constant appeals of scientists to the wealthy of this world to support research and equip scientific laboratories constituted a reflection of this state of affairs: science as a social institution did not have an institutional patron, governments had no "science policy," and the economy had no need for new theoretical knowledge. Science was independent and poor, even when one takes into account the fact that research costs then were incomparably lower than now. The very fact that scientists put so much effort into establishing the practical and moral benefits which science could bring to society clearly testifies to the fact that outside the scientific community the demand for scientific results was at that time rather small. Had such a demand existed, scientists would not have had to proclaim their usefulness so loudly.

In this situation the postulate of the disinterestedness of investigations, in the name of the moral benefits to be guaranteed by the pursuit and possession of truth, was followed in large measure *faute de mieux*. In a sense this was making a virtue of necessity, since a science that would be other than disinterested did not and could not exist. Only in the last decades of the nineteenth century did this postulate begin to be violated more and more often, and then it indeed began to function as a norm (an injunction not to do something which it is impossible to do, or an order to act in the only possible manner, does not constitute a norm). Only then was the demand for new theoretical knowledge extended beyond the scientific community; and the situation where one wrote only for one's peers began to change, as new and powerful clients for such knowledge appeared on the scientific scene.

3.

By the professionalization of science I mean the processes as a result of which the conduct of research ceased to be the private concern of individuals and was taken up as a formal occupational responsibility. In contrast with institutionalization, professionalization cannot take place without the existence of particular organizations which treat research as their mission and which are ready to supply the necessary means. Professionalization understood in this manner at first encompassed only the natural and technical sciences, while the social sciences and humanities became professionalized only in the twentieth century. One might argue that what distinguishes science today from the science of the eighteenth and nineteenth centuries is the fact that it became not only institutionalized, but also professionalized (with all the implications of this fact).

The professionalization of scientific research became possible when, as a result of the development of knowledge on the one hand, and of changes in the social structure on the other, a demand for the results of scientific work began to be felt also outside the scientific community. This would have been impossible if the development of science as a social institution had not created conditions favoring or even requiring the abandonment of the eighteenth-century model of the amateur scientist and the emergence of science as a professional career. The beginning of this process, which in our times has essentially changed the image of science as a social institution, was evident already during the last decades of the nineteenth century in the most highly developed countries in Europe and in the U.S. After the First and Second World Wars, this process assumed global dimensions and its rate has increased greatly. Countries and nations which came under the influence of scientific culture only in this century—

and this includes almost all of Asia, Africa and South America—learned to know science only in its professionalized stage, with the corresponding criteria of rationality.

With the development of research and the growth of specialization, the amateur scientist gradually became more and more anachronistic. Even a wealthy individual was less and less able to control independently the means necessary for the conduct of research. As a result, scientists became increasingly dependent on organizations willing and able to insure that the necessary means would be available. The internal development of scientific knowledge itself meant that the situation of the scientist was becoming more and more comparable to the situation of an impoverished artisan or piece worker who has only his labor to offer on the market. He has to look for a buyer of this labor force, and face all the consequences of his lack of independence; while the buyer must of course be convinced that the type of labor he purchases will prove profitable. In the nineteenth century —at least until the 1870s—such buyers virtually did not exist. Research means could be supplied only by the universities or other higher educational institutions, such as the Grandes Écoles in France; but in these institutions scientific work was still treated as a spiritual vocation rather than an occupation.

Moreover, research conducted individually, at least in the natural sciences, even when conducted by an outstanding scientist, became more and more anachronistic. (Einstein, who worked in a patent office at the beginning of his career, constitutes an exception rather than a model of how the occupational career of a researcher was shaped in the early twentieth century. But by 1913, even Einstein was working in the Kaiser-Wilhelm-Institut in Berlin, and by the 1930s in the Institute for Advanced Studies in Princeton—that is, in non-university research centers.)

The development of knowledge, its specialization, and the formation of new disciplines led to the emergence of research groups consisting as a rule of a master and his students or co-workers engaged in the same or similar problematics. Such co-operation becomes a necessary condition for the preparation of the young for independent scientific research, for which university study alone is often no longer sufficient. The formation of such groups and centers in the universities is possible, however, only if they can somehow connect the educational goals which they are supposed to pursue in the first place with research which still does not constitute a formal requirement: when they introduce some of the students to research work, and when one of the options available for the students is the professional career of a scientist (obviously on the assumption that for some reason such a career is considered attractive). In contrast to the situation in the eighteenth century, by the second half of the nineteenth century a scientific career without such an initiation was becoming less and less likely.

Certain special historical circumstances in early nineteenth-century Germany meant that the most favorable conditions for the rapid development of scientific knowledge and for the formation of the professional careers of scientists existed there.

The reforms of the German universities conducted at the beginning of the century were by no means designed with this purpose in mind. In accordance with the eighteenth-century pattern, it was assumed that a scientist was an amateur working alone, being paid for his work as a teacher and not as a researcher, and that research was a spiritual calling. At the same time, however, the principle was accepted according to which nomination to a university chair was to be an expression of recognition for outstanding scientific contributions rather than only for competence in transmitting knowledge gathered by others.

According to this assumption, the principle of habilitation and the position of *Privatdozent* were introduced, and it was from among the *Privatdozenten* that the academic senates chose the new professors. A *Privatdozent* had the right to lecture at the university, but he was not paid by the university and lived on the fees paid by the students attending his lectures. At the same time, the number of university chairs was growing, and new research laboratories and institutes were being established. A professor would both fulfil his teaching duties and at the same time direct research in his institute. These institutes became places of training for research careers, and they made the German system different from the French or British ones, which until then did not have such institutions.

As a result of the functioning of this system for several decades, around 1900 there was hardly an outstanding physiologist in the world who had not passed through the laboratory of Karl Ludwig in Leipzig, or an outstanding psychologist who had not studied with Wilhelm Wundt, also in Leipzig. The chemistry laboratories directed by Liebig, Wöhler, Berzelius, Müller, Ostwald and others probably played an even more important role, and the situation was similar in theoretical and experimental physics. University seminars played a similar role in the social sciences and humanities.[4]

Thus, when in the 1880s there was for the first time an external demand for scientists, German science could satisfy it easily, especially since it was harder to find a place in a university research institute; while at the same time conflicts began to emerge between the privileged group of ordinary professors and the extra-ordinary professors and *Privatdozenten* who were less privileged. The creation in 1909 of the *Vereinigung der ausserordentlichen Professoren*, and in 1910 of the *Verbund der deutschen Privatdozenten* testifies to these conflicts. In 1912, these two associations, which basically had the character of trade unions, united into a single organization with the revealing name of *Kartell der deutschen Nichtordinarien*. It is hard to find a better sign of the fact that by this time research was becoming a profession.

In the mid-1870s, an outstanding American astronomer, Simon Newcomb, claimed that while in Germany universities constitute the settings for scientific activity, in England and France this is the function of scientific societies. Comparing German and American universities, it was striking to Newcomb that whereas in Germany the status of the university was assured by its professors, in America, the reputation of a professor was assured by the university he worked for. Even at Harvard or Yale, professors were not expected to perform original research, but only to master the knowledge gathered by others. Newcomb believed that in the United States even if someone were to devote his life to original research and achieve success and world-wide reputation, he would not earn any more money or attract more students to his laboratory.[5]

Indeed, until the middle of the nineteenth century the American universities were to fulfill the task of providing a general liberal education to young people, while the university as a community of scholars was unknown in the States. Such universities appeared only as a result of the reforms of the second half of the nineteenth century conducted under the influence of the European (and especially German) universities. Graduate schools whose task was precisely to teach research were created as a result of these reforms.

The German example was also taken over in various ways in other countries. In France, for example, a politically influential group of scientists headed by Bertholet and Lavisse attempted to reform the French faculties on the German model in the 1880s. The organizational structure introduced in these reforms survived essentially unchanged until 1968.

All these changes in the universities and in the educational systems would not, however, have led to the professionalization of research were it not for the formation of an external demand for scientific results. This condition was met only when further technological progress became less and less possible without reliance on new theoretical knowledge. Only then did the possibilities of subordinating technology to theory that were contained *in nuce* in the modern ideal of science gain a chance of full realization.

4.

In the last decades of the nineteenth century technological innovations were resulting more rarely from

> a steady piecemeal development of improvement of existing processes; the overwhelming majority resulted from new materials, new sources of power, and above all else from the application of scientific knowledge to industry. [...] The electrical and chemical industries of the late nineteenth century were therefore not only the first industries to originate specifically in scientific discovery, but in addition they had an unprecedented impact, both in the speed with which their effects were felt and in the range of other industries they affected.[6]

This view, according to which the role of science in the social structure began to change in the last decades of the nineteenth century, is widely accepted today; but we do not always realize that as a result of these changes, by the beginning of World War I the world was already profoundly different from the world of the 1870s. And after all, these were only the first signs of what the twentieth century would bring. It is thus perhaps worthwhile to list, following Barraclough, the most important technological innovations based on the results of new scientific knowledge that were introduced in this early period.

First of all, thanks to the discoveries of Bessemer and Siemens, steel, which had previously been considered a semi-precious metal, became a readily available commodity. Its production grew from 80 thousand tons in 1850 to 28 million tons in 1900. The use of electrolytic processes and the widespread availability of electricity made possible the industrial production of aluminum, caustic soda, and electrolytic copper. (The first electrical power plant began working in 1882 in New York; the AEG was founded in Germany in 1883; the first hydroelectric power plant began working in 1890 in Colorado).

In the mid-1880s an internal combustion engine was constructed (by Daimler and Benz) as a result of which the first cars, tanks, and airplanes appeared. By 1914, more than a quarter of a million cars were riding on British roads, and in that year 265,000 cars were produced in the U.S. Airplanes and tanks were used in the First World War, and the Paris taxi drivers are reputed to have saved Paris and contributed to the victory of the Battle of the Marne by transporting tens of thousands of soldiers to the front. In 1909 Blériot flew across the Channel, producing a public sensation similar to that experienced during the first manned space flights.

Improvement in refrigeration techniques allowed for the building of refrigerated wagons and ships, which made possible the import of meat from South America and Australia to Europe. New methods for preventing food from spoiling, based on sterilization and pasteurization, made possible the storing of food and its systematic supply to growing city markets (pasteurized milk became available in 1890). New chemical and physiological knowledge contributed to changes in agriculture: industrially produced synthetic fertilizers appeared at this time.

In the area of communications, the typewriter (1873) and linotype and monotype (1887) were invented, making possible large runs of the daily press. Thanks to the use of celluloid tape, photography became popular (the first Kodak cameras for general use appeared in 1888). The telephone and microphone were invented in 1878. In 1895 the first film camera was constructed, in 1901 the first radio. The photoelectric cell appeared in 1893.

No less spectacular were the uses for new chemical knowledge. The first synthetic fibers (artificial silk, 1890) and the first plastics (bakelite, 1906) appeared at this time. Hormones and vitamins were discovered (1902). The synthesis of aniline dyes was a direct result of the development of theoretical chemistry, and in turn made possible not only the industrial production of these dyes, but—as a side effect—the identification of a large group of bacteria as well (Pasteur and Koch) and the production of vaccines against infectious diseases. The emergence of the pharmaceutical industry was a result of developments in chemistry, biochemistry and bacteriology (aspirin first went on sale in 1899, and the first antibacterial agent salvarsan was produced in 1909). With the widespread use of anæsthesia and asepsis, medical practice was being revolutionized.

Let us repeat again: these were only the beginnings of a process whose further course, stimulated by developments in modern physics, electronics, chemistry or biology, we need not relate here, but whose results are obvious today in all areas of life. Barraclough must be right when he argues that our contemporaries would feel more at home in the world of 1914 than would someone from 1914 in the 1870s. In any event, around that time technological advances became directly dependent on what was happening in the laboratories and workshops of scientists. One could say that the ideal of a science uniting the cognitive and the technological functions of knowledge achieved the possibility of full realization only then, since it was then that science itself began to create a demand for its own products.

The appearance of this external demand, which scientists had already been seeking in their appeals to those who controlled power and capital, was not only a result of the emergence of new disciplines which made possible applications such as those listed above, but also of deep changes in the economy itself.

> The expansion of the capitalist enterprise, which has been converted into a corporation, freed from the bounds of individual property, can now conform simply with the demands of technology. The introduction of new machinery, the assimilation of related branches of production, the exploitation of patents, now takes place only from the standpoint of their technical and economic suitability. The preoccupation with raising the necessary capital which plays a major role in the privately owned enterprise, limiting its power of expansion and diminishing its readiness for battle now recedes into the background.[7]

Large industrial corporations are more willing than small businesses to take on the risks connected with the dramatic changes in production, and by the same token they are more likely to establish their own industrial labs and to finance research. Thanks to this, the network of institutional interdependence between science and industry, and later between science and government, could fully develop and overcome their traditional isolation. These changes in production initiated the process of establishing research institutions outside the universities—a process which began with research on aniline dyes and vaccines and with the financing of research by large corporations, as well as by the government. So, for example, the Rockefeller Institute for Medical Research was established in 1901 (during the first five years of its existence it spent 120 million dollars on scientific research), while the Carnegie Institution of Washington (founded in 1902) was given 22 million dollars for research. The National Research Council was established by the American government during the First World War.

The European universities, at least at first, proved unable to absorb the changes to which they had contributed. The conception of a disinterested science, and the ideal of the university based on it, corresponded better to the situation of science in the eighteenth and early nineteenth centuries than to science at the turn of the twentieth century. As a result, the professionalization of science took place largely outside the universities, although the universities had created the conditions necessary for it. The often feudal character of these universities, together with fear of competition from new disciplines, especially from applied fields which were still considered less noble than pure science, and overproduction of researchers in proportion to employment possibilities even in those universities that were rapidly expanding—all this contributed to the fact that both research and personnel were being pushed outside of the universities and towards those areas where demand was rapidly expanding.

It seems that the policy of many European universities justified by the ideal of an autonomous and disinterested science had to lead to consequences quite contrary to the goals to which it appealed. Neither the autonomy nor the disinterestedness of science could be preserved by expelling from the university whatever did not fit. But if these values were to be defended at all in a changing world, then surely they were not to be defended in non-academic institutions, which by their very nature directed their scientific knowledge to a different audience, applied different norms to their activities, and—what is equally important—had no traditions linked to the attempts to realize these values.

In any case, by the end of the nineteenth century all the necessary conditions were met for the transformation of the links between science and industry and government, which until then had been sporadic, into stable and institutionalized relations; and the process of the integration of science with other social systems—above all with the economy—could begin. This was also the beginning of Big Science, in which basic research cannot be distinguished from such initiatives as the Manhattan project or the Apollo program. The enormous bang of the explosion of the first atomic bomb in August 1945 was not so much the beginning of a new period in the social history of science as an event drawing the attention of the whole world to the changes in science that had taken place during the several decades of its industrialization and professionalization.

Let us add that the formation of mass societies, great city conglomerates, the inclusion of the masses in political life, the formation of mass political parties, and last but not least the emergence of totalitarian systems, created in turn a need for social knowledge which could be used to manipulate people and symbols as if they were things. As a result, the social sciences and humanities underwent a similar process of professionalization and of growing dependence on economics, state and group ideologies, and politics.

The processes discussed here began without arousing much notice, but they continued at such a rate that after the First World War and especially after the Second, they became a reality which had a fundamental impact on scientists' self-knowledge and on the public view of science. The autonomy of science, understood as its independence from the global social structure, corresponded less and less to actual situations. The belief in such an autonomy became an element of false consciousness. The ideal of the unilateral and beneficent influence of science on social life was losing its justification in the real world.

CHAPTER V

THE SOURCES OF THE CRISIS
OF THE MODERN IDEAL OF SCIENCE

1.

It is impossible to analyze here all the consequences of the processes we have just described; I will limit my discussion only to their most important effects on science as a social institution, which also contributed to the crisis of its modern ideal. The second source of this crisis—a purely cognitive one—will be treated in the following sections of this chapter.

First, the professionalization of research significantly changed the nature of the audience to which scientific work was addressed. Until then, it was only other scientists, the potential authors of similar communications, who constituted the intended audience of the contributions of scientists. The scientist worked primarily for other scientists, and as a researcher he communicated only with them (although not as a teacher or writer of popular accounts of science). Their evaluations of scientific results were important, since these evaluations were decisive in according the scientist the social reputation to which he aspired. This situation was not affected by whether or not the scientist believed that the knowledge he was producing might also be valuable beyond the circle of his fellow researchers. The lack of institutional connections between science and other social systems allowed him to believe that he was supplying only disinterested information about the world, information which might possibly be of practical use, but whose value did not lie primarily in such practical utilization. In other words, a scientist could then believe that he was a purveyor of pure truth.

From this perspective it was unthinkable to treat science as a "productive force," as a tool for the realization of goals other than cognition, or to consider the training of specialists in terms of the development of a work force. "For centuries universities have been concerned with individual men and not with man-power. No one, except as a joke, talks about classical man-power or philosophical man-power."[1]

When scientific research becomes a professional duty which the scientist owes to the organization that funds his research, the situation changes dramatically. The scientist then becomes either an employee working under the control of a supervisor, or an individual free-lancer living on income from contracts from institutions conducting research, or a manager who organizes research. These roles are obviously not mutually exclusive.[2]

As a researcher, the scientist will obviously continue to strive for recognition from the community of his peers, but even in the best of cases, their evaluations will no longer be his only frame of reference. Moreover, changes also occur in the criteria for making such evaluations and in their hierarchy within the scientific community. Truth ceases to be acknowledged as an autotelic value for everyone. True knowledge comes to be valued above all for its utility; and utility, in contrast to truth, is a matter of degree; moreover, it is relative to the addressee and to the circumstances. The addressee is now the institution which employs the scientist and finances his research. And even if truth and utility were not as a rule contradictory values, even if one believed that only true knowledge is useful—that is, even if we exclude the drastic cases where falsehoods or lies have been adopted as useful, and when scientific knowledge has been subordinated to narrow political or ideological interests—still, in a given situation and for a specific addressee, *not all true knowledge is always useful*.

The fact that not all true knowledge is useful leads to a situation in which research in professional science becomes a subject of planning and of "science policy"; and in which those who believe that it is good to know, even when knowledge brings with it no satisfaction beyond the quest for truth, find themselves in a more and more difficult situation.[3] This is so because science ceases to be treated exclusively as a supplier of disinterested information and becomes a producer of prescriptions for manipulating objects, symbols, and people. This characteristic becomes a feature distinguishing science from other spheres of cognitive activity. It appears to be far more adequate as a distinguishing characteristic of modern science than all methodological criteria of demarcation.

Secondly, while in the past the choice of a research topic was dictated exclusively or primarily by the interests of the researcher who himself remained in control of the information resulting from his work, today the choice of a subject becomes more and more dependent on the needs of the market, and even more directly on the recognition of these needs by the organizations financing research. The publication of results begins to depend more and more on various extra-scientific considerations, and the decision whether or not to publish often does not depend on the scientist alone.

If previously a researcher was paid regardless of the utilitarian value of his accomplishments, and often regardless of the results of his investigations, now the results he achieves can be bought or sold, and more and more often he must play the role of an expert paid by industry, the army, or the various governmental agencies. He becomes an expert whose function is not so much to choose goals as to advise about the means of realizing goals that have already been chosen, and which he cannot influence to any significant degree. Moreover, such choices are often made in secret, and then they are not even subject to discussion among specialists.[4]

Clearly, in this situation of conflicting loyalties, the actual behavior of scientists deviates more and more from the norms of the scientific ethos formed in earlier times. The moral principles accepted until now in the scientific community are severely tried. What can one expect when the scientist participates in two groups: when a claim is made in the first, its members will wonder only whether or not it is true; while when a claim is made in the second, the others will first consider why it was made, and only later perhaps worry about whether or not it was true?

Thirdly, the professionalization and industrialization of scientific research threatens the traditional system of control over research work and the publication of results. This is so because on the one hand, the products of scientific research are in fact understood only by other scientists and only other scientists can evaluate them substantively; while on the other hand, the actual dependence of science on other social systems, above all on the economy and politics, means that the researcher is less dependent on the evaluations of his peers and thus is more likely not to take their views into consideration to the same extent as before. This is relatively easy given that even under the best conditions, the system of control and evaluation of scientific results can never be fully formalized, because one of the basic characteristics of scientific creativity was and remains the ability to break away from tradition, and because substantive issues in science cannot be solved by the vote of a majority, even a majority of competent specialists. The history of Lysenkoism shows what can happen when the normal mechanism for denouncing quacks is blocked by other social mechanisms.

Fourthly, in these circumstances the situation of the scientist becomes morally ambiguous in two ways: first because absolute adherence to the scientific ethos becomes difficult; and secondly, because of the consequences which could follow from a strict adherence to these norms.

As science becomes increasingly subordinated to the economy and politics it ceases to be the same social system as before. It becomes rich but it loses its autonomy. It gains great influence, but it begins to function according to rules other than those which until then had been constitutive. How can a scientist meet the demand that his results be subject to the criticism of all competent researchers when these results are kept secret, and when sometimes even the very subject of the research is not public knowledge? How can this be expected of scientists when they no longer exercise full legal control over the knowledge they produce? How can a scientist take account only of the evaluations of his peers when the most fundamental aspects of being able to conduct research depend also (and sometimes above all) on evaluations from outside the community of researchers and on evaluations based on criteria different from those used by disinterested and competent judges? How can one expect that in industrialized or professionalized science, not to mention militarized science, a scientist's behavior will be guided by the principle formulated once by Descartes, who vowed not to pursue investigations "which can only be useful to some by being harmful to others"?[5]

We can multiply these kinds of questions in relation to all the norms once considered constitutive of science. Thus one can doubt whether the norms of the scientific ethos analyzed by sociologists still give a valid account of the real functioning of science as a social institution, or whether they refer rather to an ideal which corresponds less and less to the actual state of affairs. And one can then wonder whether the continued characterization of science in terms of these norms is not just a means of perpetuating a particular mythology. This is so not because scientists do not obey the norms—violations have always occurred—but because we are now dealing with a different social system, which strives to realize an ideal of knowledge different from the one once constituted by traditional scientific norms.

On the other hand, a scientist who would continue to obey all these norms in today's system of professionalized science would not be guaranteed the morally superior position which he would have enjoyed earlier, in times of "clear conscience." The more cognitive activity is subordinated to practical goals, even independently of the individual motives behind each particular decision, the more deceptive will be the conviction that such activity can protect the scientist from the moral conflicts inherent in all human activity.

A change in the social status of science such that traditional ethics cannot survive a confrontation with reality, and that its ideology changes into false consciousness, is not only a source of personal drama for many scientists (among them, for example, Albert Einstein), but forces us all to think about science in new categories.

If an innocent formula like $E=mc^2$ can provide a basis for the production of weapons of mass destruction, then the moral conflict of the scientist goes beyond the fact that in signing a letter to President Roosevelt he has contributed to this application of his discovery. It is a far deeper problem, since it transcends the realm of political mistakes—the authors of the letter were motivated by fear that Hitler's Germany might produce an atomic bomb before the allies. The issue is not one of an individual violation of the Cartesian principle cited above, but rather the far more basic issue of whether science in general, even in its most disinterested form, can remain consistent with this Cartesian principle. In a world of industrialized and professionalized science even the purest striving for truth ceases to be a morally innocent activity free of ethical conflicts. Not only is it no longer possible to maintain the view that scientific development constitutes an unadulterated blessing for humanity, but even the weaker conviction that this development is morally neutral ceases to be plausible. The most that can be said is that science is morally ambivalent, and ambivalence is certainly not the same thing as neutrality.

Given the shocks of recent years, the moral ambivalence of the very search for truth has become an essential condition of scientists' self-knowledge, at least in the case of those scientists who do not limit their moral responsibility to a respect for the rules of the game as codified in the methodology, but extend it to the consequences of their participation in the game itself. The difference between the indictments in the trial of Galileo and the trial of Oppenheimer testifies to the essentially different context in which the moral responsibility of scientists must be considered today. None of the benefits which science has brought and continues to bring humanity are able to obviate the fact that science also plays a role in all the dangers threatening our civilization and our culture. It is in this sense that it is not morally neutral but ambivalent.

Scientists who today try to reach agreement, not only not to publish the results of research in certain areas (which was the essence of Szilard's appeal of 1938), but also to impose limits on the conduct of investigations themselves (as was done by the participants in the Asilomar conference in California), no longer believe that discovering the truth is always and under all conditions beneficial to humanity; nor do they believe that their moral responsibility is limited to the following of methodological rules in conducting their research. When the rationality of the effects of the development of science ceases to be unproblematic from the perspective of the cultural values that are to be realized, the philosophical reflection on science also ceases to concern itself exclusively with questions of the methods to be used in gaining true knowledge. "The old image of science as the 'endless frontier,' on which a whole generation has been brought up, seems to be giving way in some quarters to the notion of science as the suspected frontier. For whether one likes it or not, the disputes concerning

the wisdom or danger of placing 'limits on scientific inquiry' may have
been inevitable and were perhaps overdue."[6]

2.

The crisis in the modern ideal of science was, however, brought
about not only by the processes discussed above. It emerged also (and
insofar as the consciousness of this fact among scientists and philosophers
is concerned, it emerged mainly) from the very cognitive successes of
science, since they undermined their own epistemological bases and
brought about, as before in the sixteenth century, yet another revision of
the global image of the world and of the status of the knowing subject.

It turned out that despite its initial assumptions, the cognitive effort
based on the modern ideal of science undermined more and more seriously
the ideal of an autonomous knowing subject who, observing the world
from a privileged, almost divine position, was to be able—like Laplace's
demon—to gain knowledge which would be always and everywhere valid,
knowledge whose content and value were to be totally independent of the
investigator's own place in the world of nature and culture. In both the
social and the natural sciences, man as a knowing subject was increasingly
incorporated into the world of nature and society, as his cognitive possibilities of learning about this world were increasingly relativized. By the
same token, the claim that science leads to universally valid knowledge,
since its content depends neither on the individuality of the subject, nor on
his physical or biological characteristics as a member of a species, nor on
his location in history, became questionable. The cognitive autonomy of
the subject, his ability to gain knowledge unmediated by his characteristics
as a subject—valid for all subjects regardless of their material constitution
or historical placement—is now being questioned not only for theological
reasons, but for social and biological reasons which have been disclosed by
the very development of knowledge in physics, biology, and the social
sciences. By the same token, the problem of the relationship between
nature and culture has become a central issue which must be addressed if
we are to understand the relativization which has also occurred in the
context of the analysis of the cognitive value of science. Cartesian dualism
guaranteed the autonomy and rationality of the subject, while the radical-
empiricist idea of knowledge asserted that the theoretical reason is in direct
contact with apodictic empirical facts, so that the theories it constructs are
based on a completely independent and objective foundation; however, both
of these conceptions turn out to be equally incapable of explaining the

process of human cognition. The Kantian synthesis, without the concept of a transcendental subject which cannot be defended on scientific grounds, will force us to ask questions about the biological and/or historical conditioning of the forms and categories of cognition which from the Kantian perspective were to be treated as *a priori*.

In this situation, the philosophical reflection on science encountered the problem of whether it is at all possible to justify the claim that science can supply universally valid knowledge, that is, knowledge which would not be influenced at all by the biological characteristics of the knowing subject, by his conceptual apparatus used in articulating knowledge and representing reality, by the language in which this knowledge is formulated, by the historical conditions under which science is practiced, or by the culture of which it is a part. Does the very recognition of such conditioning as no longer an accidental bias which to some extent could be eliminated but rather as an unavoidable and ever-present determining factor, allow us to treat scientific knowledge as the embodiment of an eternal, unchanging human rationality, and as knowledge valid in all possible worlds and for all possible subjects? And if not, can science, together with its logical foundations, be treated as something other than a means of biological adaptation of the species (adaptation which cannot be said to be either true or false), or as an instrument for the effective manipulation of objects and symbols (an instrument which also cannot be evaluated in terms of the categories of truth and falsehood), or as a system of statements accepted on the basis of particular linguistic conventions? And what is to become of the cognitive autonomy of the subject and its rationality if the classical concept of truth were to be replaced in science with the concept of the acceptability of statements in terms of criteria other than their substantial relation to reality (such decisive criteria are lacking even though the concept of truth itself, as Tarski has shown, can be properly defined)— criteria such as instrumental effectiveness in manipulating objects, increasing the chances of the biological survival of the species, a consensus among specialists who enjoy social trust, or logical coherence according to conventions either consciously chosen (as in artificial languages) or adopted by custom (as in natural languages).

In a word, must we replace the ideal of an experimental science, asking questions of nature and reading its answers in the language in which the book of nature was written, with an ideal of instrumental science? The term "instrumental" is used here not to refer to the fact that science uses instruments, as discussed above, but in the sense that science itself becomes an instrument to serve the realization of goals external to itself, and can be evaluated in terms of the criteria adequate to these goals.

The basis for the formulation of questions such as these can be found in almost all the contemporary achievements of science, which, paradoxically, emerged from the cognitive programme formulated on the basis of the modern ideal of science together with its conception of the autonomous knowing subject.

3.

According to the Newtonian synthesis, the crowning achievement of the scientific revolution of the sixteenth and seventeenth centuries, the world was composed of matter, built of an infinite number of discrete, insoluble and unchanging corpuscles (atoms); and of motion, which does not change these corpuscles but relocates them from place to place in an infinite and uniform space in which the atoms and bodies made of atoms move. Universal gravitation kept this world together, acting both immediately and at a distance.

This was the brilliant synthesis of two tendencies: one derived from the ancient atomist philosophy of Democritus, Epicurus, and Lucretius, whose most outstanding modern representatives were Gassendi and Boyle; and the other, the mathematization of nature of Galileo and Descartes. "For him [Newton], just as for Boyle, the book of nature is written in corpuscular characters and words. But just as for Galileo and Descartes, it is a purely mathematical syntax that binds them together and gives its meaning to the text of the book."[7]

However, the creation of non-Euclidean geometries by Lobachevsky, Riemann and Bolyai gave rise to a bothersome question: what is the status of mathematical knowledge if, on the one hand, one cannot decide empirically which of the various geometries corresponds to the actual properties of physical space (that is, what really is this syntax); while on the other hand, because various geometries are possible, one cannot continue to maintain that mathematics in general and geometry in particular constitute synthetic *a priori* knowledge in the Kantian sense of the term (that is, that only one such syntax is possible)? What does it mean to say that the book of nature is written in a language of straight lines, circles and triangles, or in a corpuscular language whose syntax is mathematical, if in either case the language can be read in a variety of ways and there is no privileged manner of reading it, or at least no such manner is encoded in the human mind.

The dispute between the formalist and the intuitionist directions in mathematics, which began with Frege, Peano and Hilbert and continues to this day, concerns in essence precisely this question: does the work of the mathematician consist only in logical operations on symbols which are not limited by anything other than accepted conventions, or is it in addition an attempt to deal with a reality which exists independently of the subject and which at least limits the freedom to create conventions? In the first case, the truth of mathematical theorems is relativized to linguistic conventions, and the question of why we choose certain conventions and not others cannot be resolved other than by an appeal to the convenience of a given choice. In the second case, opposition to such relativization and to regarding logic and mathematics exclusively as formal linguistic rules leads to a subjectivist epistemology in which "to exist means to be constructed" (L. E. J. Brouwer), and the truth of statements about existence is guaranteed by intuition. Whichever of these answers we accept, we can no longer say that in doing mathematics we are reading the book of nature as it was written.

In a broader perspective, this problem is not limited to mathematics only, since every theory can be presented as a formalized deductive system, open to the question of whether its truth is guaranteed either by the syntactic and semantic rules of the language in which it is formulated or by its instrumental effectiveness. In the first case, all changes in our knowledge could be presented simply as changes in the language used to describe the world, and every theory which experience puts in question could be saved by an appropriate linguistic reinterpretation. This view is the foundation of all the versions of radical conventionalism and the source of the Duhem-Quine thesis which excludes the possibility of a crucial experiment in science (we will discuss this in detail in chapter VII). In the second case, we end up with the view that whether knowledge is true or not is decided by its instrumental applications. This view is the cornerstone of various versions of instrumentalism and operationalism. The problems we have been discussing, however, were the result of developments not only in logic and mathematics but also in the experimental sciences.

Everyone knows of the resistance which Newton's theory encountered among the adherents to Cartesian physics, according to which extension and motion were the only categories, and the world was a plenum rather than a vacuum. Perhaps no one has described this dispute as well as Voltaire: "A Frenchman who comes to London [...] has left in Paris a full world, and here he finds it empty. In Paris everyone sees the universe as consisting of vortices of subtle matter, while in London nobody notices anything of the kind [...] Among our Cartesians everything happens as a result of a stimulus which nobody understands; for Mr. Newton everything is given by an attraction whose cause is also unknown."[8]

Eventually, Cartesian physics failed to give birth to a science that could compete with Newtonian physics. Fifty years after the publication of the *Principia* (1687), the leading continental physicists and mathematicians—Maupertuis, Clairaut, d'Alembert, Euler, Lagrange and Laplace—were all convinced of the Newtonian theory and were busy improving and further specifying the structure of the Newtonian world. As a result, towards the end of the eighteenth century, thanks to Lagrange's *Mécanique Analytique* and Laplace's *Traité de mécanique céleste*, the Newtonian system seemed to have achieved its final perfection, and future generations of scientists learned not to notice the problems and contradictions which just a short time earlier had been a subject of heated controversy. What is more: the success of Newtonian physics meant that the specific features of this theory came to be considered as the necessary characteristics of all scientific knowledge, and all the new disciplines which have emerged in the eighteenth century, including the sciences of man and of society, have attempted to conform to the Newtonian model of empirico-deductive knowledge and to follow his famous *Regulæ philosophandi* (which were of course interpreted in a number of different ways). In any case, Newtonian mechanics was supplemented by a certain global vision of the world, generally known as mechanicism.

I do not intend here to follow in detail the arguments surrounding Newtonian physics (concerning mostly the nature of time, space, gravitation, or momentary action at a distance), nor to discuss its successes in explaining astronomical, mechanical, acoustical, thermal, or optical phenomena, nor to analyze the often uncritical attempts to apply its principles and methods to investigate the domains of phenomena completely different from those in which Newtonian physics was so successful. I am interested here only in the concept of the knowing subject linked with this theory, or more precisely in the issue of the physical limitations on his investigative activities within the mechanistic vision of the world.

A representative view of this matter can be found in the famous fragment of Laplace: "The intellect which at a given moment would know all the forces active in nature and which would in addition be voluminous enough to subject this data to analysis, would in one glance encompass motions of the largest bodies in the universe and of the lightest atoms. Nothing would be certain for it. The future, just like the past would be spread in front of its eyes. The human mind gives us a ceratin pale image of this intellect thanks to the perfection that humans could give to astronomy. Discoveries in the areas of mechanics and geometry, together with the discovery of universal gravitation, allowed him to contain in the same analytical expressions both the past and future states of the universe.

Applying the same method to other objects of cognition, he arrived at the reduction of the observed phenomena to general laws and to the possibility of predicting phenomena which specific circumstances could cause. All his intellectual efforts to know the truth bring him constantly back to the idea of the intellect which we described above; and yet the distance between them remains infinite. This is a tendency specific to humankind."

Thus a tendency specific to the human species as a knowing subject, a tendency realized through the cooperation of individual subjects over generations, amounts to a constant approach to the perfection of the demon who in one glance can encompass all of the past and future, and whose cognitive abilities are not limited by any physical conditions, or by any particular relations obtaining between him and the object of his cognition. For a human knowing subject, approaching this ideal consists in the improvement of his capacities for experimentation and theoretical analysis; but this process of achieving perfection which is not physically limited in any way can proceed infinitely, and in this sense Laplace's demon is a legitimate model of the real knowing subject.

Classical mechanics allows for such a conception of the knowing subject first of all because the unequivocal character of mechanical laws (their symmetry with respect to time) implies that every state of the universe as a whole, and every state of an isolated system, contains full information about the past and the future of that system. There are no signs of the past which become obliterated (for example because some objective possibilities have not been realized); there are no present states which could not unequivocally define future states; and there are no possibilities which would not eventually be realized. Since nothing prevents the subject from accelerating its empirical reconnaissance in the world or from reasoning about it, there are no physical limitations to his ability to predict the future.[9]

Secondly, classical mechanics allows for such a possibility since it makes it perfectly possible to assume that all information about the past can be available at any moment. All actions, and thus also all transfers of information, occur instantaneously, and there is no place for something like the subject's temporal horizon, which for purely physical reasons defines the boundaries from beyond which information could not be obtained.

Thirdly, classical mechanics allows us to ignore completely the problem of the costs involved in gaining information about the state of the investigated system, if it is assumed that these costs can be minimized arbitrarily, that is, that they depend exclusively on the subtlety of our experimental methods. The implicit assumption of classical mechanics about the relations between action and information allows us to exclude from consideration the influence of the experimenting subject on the phenomenon under investigation, especially when this subject is seen as a human species historically perfecting its cognitive capacities.

Incidentally, for the same reason, the epistemology supplementing classical mechanics does not and cannot treat the problem of uncontrolled disturbances in the investigated system, which could, for example, threaten a catastrophe for the species. This is one of the reasons why "scientific activity is analyzed in this era only as a potentially more and more powerful stabilizer of the conditions of life necessary for the population, but not as a potential source of real threat to these conditions."[10]

Classical mechanics, in other words, does not impose any physical constraints on the abilities of the human subject in approaching the ideal of Laplace's demon. This trend "specific to humans" is not limited by the nomological structure of the world, which allows for a full transmission of information through time, nor by the time needed for the gathering and processing of information, nor by disturbances in the investigated system caused by the process of investigation itself. Moreover, classical mechanics does not allow for the problem of population threat emerging from the experimental positing of questions to nature, and thus also does not allow for possible restrictions which can be placed on such activities. The optimistic vision of "knowledge without limits" can appeal to this trend as a justification implicit in nature itself. There are no physical reasons preventing the subject from perpetually perfecting his investigative activities.

Ontologically the subject belongs to the world he is investigating, but as a knowing subject he can treat himself as if he were an external observer limited by no physical constraints and completely abstracted from the investigated world. Thanks to this he is able to gain objective knowledge; he can be autonomous with respect to his physical environment.

Contemporary physics questions this idea of the subject. The theory of relativity presupposes the existence of a temporal horizon from beyond which the subject cannot possess any information, and it assumes that any transfer of information requires time, and does not take place instantaneously. This obviously limits the possibilities of both postdiction and prediction, even for Laplace's demon.

These possibilities are much more radically limited by the rejection of the concept of the *unequivocally* deterministic character of the fundamental laws of physics, and the adoption of Heisenberg's indeterminacy principle, which states that every act of measurement or observation disturbs in an uncontrolled manner the state of the system under investigation, and that the possibilities of perfecting the investigative system are by no means unlimited: its limits are determined by a physical constant, known as the quantum of action. It follows that there is a physical limit to the human approach to the ideal represented by Laplace's demon. We always pay a price for obtaining information: we disturb the state of the system. The subject's knowledge is not independent of his physical characteristics as an object dynamically linked to the system. The uncontrolled character of this disturbance, which can be described only in statistical terms, means that knowledge of the system has to take into account not only the physical characteristics of the subject and its measuring apparatus, but it can never be complete in the way in which the demon's knowledge was complete. In a word, on the basis of modern physics, Laplace's demon can no longer be treated as an ideal model of the knowing subject, an ideal which the human species was to approach ever more closely in the course of an infinite historical development. The cognitive possibilities of the subject are now rendered relative to his physical nature, and the content of his knowledge of the world cannot be independent of the fact that he is himself a physical object belonging to the world he is investigating. There are physical limitations to the subject's autonomy with respect to the world he is investigating.

The achievements of the sciences of man—of biology, neurophysiology, linguistics and cultural anthropology—were no less important for the problem we are considering here than developments in physics. Obviously, we cannot discuss them here in detail. Accordingly, it is necessary to conclude with the statement that all these developments force us to treat the knowing subject not only as a physical object, but also as a biological and social object; and the epistemological consequences for the character of scientific knowledge which this treatment entails are analogical to those described above.

The formulation of the theory of evolution and the later development of genetics, revealing the biological conditioning of all the intellectual abilities of man, led to questions of whether and to what extent the subject's knowledge can be always and everywhere valid independently of the biological characteristics of the species. By the same token, they paved the way for the conviction that human intellectual capacities and their products, like the other capacities of living organisms, constitute a particular means of human adaptation to the environment, and that their value is limited to this function.

Sociology (including the sociology of knowledge) and anthropology, breaking through the Eurocentrism characteristic of the European humanities in the nineteenth century, have forced us, as a result of investigations of other cultures, to look "from the outside" at our own culture as well. By the same token, they have made it problematic for those who participate in it, since this culture could no longer be treated as the only possible one, the only rational one, or the only one corresponding to "human nature." Instead, one had to think about its own dependence on surrounding circumstances, and to relativize its values, which until then had been seen as unproblematic. From this point of view, the development of linguistics and various types of language studies has had similar consequences. Constantly encountering the problem of the dependence of the human vision of the world on the linguistic apparatus by means of which reality is articulated, the philosophical underpinnings of the sciences of language oscillate between biological and cultural relativism. It is probably not a coincidence that the idea of radical conventionalism with reference to the artificial languages of scientific theories appeared at roughly the same time as the Sapir-Whorf hypothesis concerning natural languages.

The rejection in biology and sociology of the eighteenth-century concept of an unchanging human nature led to the drawing of distinctions between natural and artificial elements in culture; and the opposition between culture and nature became the paradigm of philosophical reflection about all forms of human activity and its products.[11] As a result, philosophical reflection about the nature and value of scientific knowledge has also had to face the problem of the biological and cultural conditioning of the knowing subject. The admission that this conditioning is not accidental, and that it cannot be eliminated, but remains an inseparable feature of all processes of cognition, has already undermined the idea of the autonomous knowing subject and with it the modern ideal of science based on this concept of the subject.

This new perspective forces us to ask the question of whether and how any science, whether considered as a means of adaptation for a particular biological species, or as a part of a specific, historically conditioned culture, can justify its pretensions to universal validity. It forces us to consider the influence of both biological equipment and of socially inherited culture on the process of learning about the world and on the rules governing the acceptance and rejection of empirical claims, explanations of phenomena, and the construction of theories. In a word, this perspective has disintegrated the idea of the rational knowing subject which supported the modern ideal of science and allowed us to treat scientific knowledge as the embodiment of the rationality inherent in human nature. The development of scientific knowledge itself led to a questioning of the autonomy of the knowing subject with respect to all external factors, and this in turn had to result either in the impossibility of treating the development of science as a purely rational process, as the expression of the evolution of Reason taking place according to its own immanent laws of one type of logic or another, or it had to result in a revision of the concept of rationality itself, relativizing it either with respect to biology or with respect to culture. This is the context of contemporary discussions of the rationality of science and its development.

Thus, while the professionalization and industrialization of science undermined the autonomy of science as an institution and made its rationality problematic with respect to the cultural values which gave birth to science and of which science is a part, the very development of scientific knowledge undermined this ideal as if from the other side: it questioned the idea of an immanent human rationality implied by this ideal, and it promoted the formation of a new kind of scientific self-knowledge, a self-knowledge attempting to trace the historical, sociological, and cultural conditioning of the development of knowledge.

Although it would be difficult to claim any direct connection between these two processes, it seems that without great oversimplification we can claim that more than a coincidence linked together the actual undermining of the autonomy of science as a social institution and the philosophical controversies about the autonomy of the subject. These were the two—logically irreducible—sides of the crisis of the modern ideal of science, of an ideal which connected the epistemological idea of the autonomy and rationality of the subject with the postulate of the autonomy of science as a social institution.

Both of these processes together led to the crisis of nineteenth-century scientism, the dominant element of the scientists' self-image based on the modern ideal of science.

4.

If the achievements useful for humanity move your hearts, if you are amazed by the surprises of the electric telegraph, anaesthesia, the daguerreotype, and of many other excellent discoveries, if you care about your country's participation in the development of these marvellous achievements, then please take an interest in the holy places which are appropriately called laboratories. Demand that their number be multiplied and that they be better equipped. These are the temples of the future, of wealth and welfare. It is there that humanity matures, gains strength and perfects itself. It learns there to read the works of nature, progress and universal harmony, while its own works are all too often those of barbarism, fanaticism and destruction.[12]

With these words Louis Pasteur appealed to his compatriots in 1867, pleading for the financing of scientific research. On a different occasion, however, Pasteur also wrote:

> The cultivation of the sciences in their most perfect form is perhaps more necessary for the moral state of the nation than for its material welfare. Great discoveries and meditations, inspiring art, science, and literature, in a word, all disinterested works of the mind in all disciplines, and the educational establishments popularizing these achievements, instill in all of society the philosophical and scientific spirit which subordinates everything to the demands of reason, condemns ignorance and eliminates superstition. They raise the intellectual level and moral consciousness; even the idea of God spreads and flourishes thanks to them.[13]

Henri Poincaré argued similarly,

> The scientist should not waste his time on the achievement of practical goals. He will surely reach such goals, but this must be marginal with respect to his principal activity. He should never forget that the specific object he is investigating is part of a whole which is infinitely greater than this object; love for this whole and an interest in it should constitute the only motives of the actions of the scientist. Science has marvellous applications, but a science in which applications were the only aim would no longer be science but only a kitchen. There is no science other than disinterested science.[14]

These statements date from a period when the social status of science was already undergoing major changes, but when scientists and philosophers were not yet aware of these changes. I have cited them here since they illustrate two not completely compatible self-images of scientists founded on the basis of its modern ideal: one of these treats science as a *system of beliefs*, a supplier of truth, and assumes that for moral reasons "all truth discovered on the surface of the earth is beneficial for all of humanity"; while the second treats science as a *means of manipulating objects*, a source of technological prescriptions, and assumes that science is a means of achieving wealth, welfare, economic and state power, or national pride— values of a different order than truth and moral good.

It appears that these two seemingly contradictory images constituted, at least for a time, a consistent whole which even into the first decades of this century was understood as an adequate representation of science, a whole which is usually referred to as scientism. But when the changes discussed earlier were consciously recognized, this whole gradually disintegrated and became more and more a form of false consciousness.

The term "scientism," like other designations of philosophical trends, is used in philosophical literature in many different ways. Some authors tend to identify scientism with a broadly understood positivist trend in European philosophy, from Hume to whatever is left of logical positivism today.[15] Others link scientism with beliefs about the scientific method of nineteenth-century scientists and philosophers who were convinced that "science is the only trustworthy source of knowledge about reality."[16] Friedrich A. Hayek, and others in his wake, define scientism as a "slavish imitation" or "unjustified transfer" of the methods and language of the natural sciences to the study of man and society.[17] In this manner they oppose the notion of the methodological unity of all scientific knowledge; but the question of where the justified transfer of this method ends, and unjustified transfer begins, is answered by each of these authors differently. Popper modifies Hayek's definition, claiming that scientism is the imitation of what is generally, but erroneously considered the method of science, from which we can conclude that there is one real scientific method; the problem results from the fact that the methodological views of many scientists do not give an adequate account of the methods they really use. It seems to follow clearly that propagating this one adequate method would no longer be called scientism.[18]

Popper's correction is obviously designed to insure that the term not be applied to his own position, nor does anyone else admit nowadays to being a proponent of scientism. As the above definitions show, the term scientism is now used only in a pejorative sense, but less than a hundred years ago A. Roy called his own position scientistic and wrote: "My conclusions are rationalist and intellectualist. I do indeed believe that rationalism, since it constitutes an absolute justification of science, should be based on science and not go beyond it."[19]

I do not intend to engage in arguments about what is an adequate definition of the term "scientism," since it is well known that such arguments are usually fruitless. What I would like to do, however, is to bring to light a thesis of particular importance in controversies about the rationality of science and its development, and one which has doubtless constituted one of the characteristics of a scientistic position in all its variants. The question of whether the acceptance of this thesis can be taken as a feature differentiating scientism from other positions is in this context quite irrelevant.

By "scientism" I understand a certain set of theses concerning *the social value of science and of the scientific method*. In this sense scientism is not equivalent to the attitude of a naturalist as such, or to any definite ideas—correct or mistaken—about the scientific method, but with a particular opinion valorizing the role of science in culture. This might be the view of a layman as well as a scientist. Most generally speaking, it is the opinion that science is an unproblematic good.

It is true that in the eighteenth and nineteenth centuries scientism was generally linked with the acceptance of a radical empiricist view of the scientific method. I believe, however, that what constitutes a distinctive feature of all scientistic positions it is not the identification of a specific scientific method deemed rational, but the conviction that because of the rational nature of science any scientific method can serve as an ethical code of science.

Only on the basis of this understanding of scientism can we discuss possibly its contemporary, twentieth-century continuations. If we were to construe this concept in such a way that scientism would be linked with a definite idea of the scientific method—for example with radical empiricism or positivism—then we would have to admit that scientism has disintegrated as a tenable position in philosophy of science in our time. Given such a terminological decision it would be better not to apply this term to any of the contemporary positions in philosophy.

Because scientism treats science as an unproblematic good, it postulates that society should fully accept a system of values whose realization will create the best possible conditions for the development of science, or more precisely: for the realization of one of its ideals. This was the source of the aggressive character of scientistic thought, which demanded the subordination of all social life to one or another version of the scientific method.

This tendency found expression both in the Enlightenment faith that propagating the scientific method in social life might constitute a universal remedy allowing for the rational solution of all social conflicts and leading to progress, and in the social strategy according to which scientists were supposed to focus all their efforts on the development of knowledge without getting mixed up in politics, while attempting to persuade the wealthy of this world that they should support the development of science with all available means and without interfering in its internal life. In the days when scientists had not yet made the voyage from the *New Atlantis* to *Brave New World* and *1984*, many of them found this ideology attractive.

This ideology, however, was also founded on the idea of the autonomous knowing subject which we discussed above, a subject who thanks to the right method could approach universally valid knowledge, and on the corresponding vision of science as a social institution. The undermining of these premises led to the crisis of the scientistic ideology. In other words, in a situation in which scientific activity was in no way institutionally linked with the economy and politics, the scientists' reflection about their own activities had to focus primarily on epistemological and methodological issues, while they could treat the institutional autonomy of science as a normal, almost natural state of affairs. The special historical situation under which scientists lived and worked until the last decades of the nineteenth century could seem to them the only possible one. This conviction, considered as universally valid and as a reflection of the very essence of scientific activity, turned into false consciousness when, on the one hand, these special historical conditions no longer obtained, and on the other, when the development of science itself undermined the idea of an autonomous knowing subject.

If one does not doubt that the knowing subject can be completely autonomous in his attempts to learn about the world, that is, that he is able to exercise his cognition unaffected by his individual, biological or historical characteristics, and thanks to this autonomy or rationality is able to discover true knowledge; and if at the same time we believe that true knowledge is always beneficial, and that by accelerating the development of knowledge one is unequivocally and undoubtedly making a moral and material contribution to the well-being of others; then the responsibilities of scientists can be exhaustively defined in terms of guarding the autonomy of science from all external, doctrinal or political interference, improving the method thanks to which the autonomy of the subject can be realized, and obeying the rules of this method in their research.

By accepting these premises, the scientist could say to himself, "My ethics is my methodology. So long as I do not sin against the rules of the game of science as they are codified in this methodology, I am also meeting my professional responsibility as a scientist and my ethical responsibility as a human being." He could then also believe that the profession of a scientist assured its members a privileged moral situation. Since the goal of the scientist's activity is morally sanctioned *a priori* and constitutes unequivocal good, then the moral duty of the scientist is simply to seek to achieve this goal by the proper means. Reflecting on his activities as a scientist, he has to consider mainly, if not exclusively, the issue of "what are the directives of investigative activity, and under what conditions can we achieve truth by following them?"[20] By attempting to realize values which were seen as important not only here and now but always and everywhere, he did not even have to experience the moral conflict of choosing between loyalty to his nation and loyalty to humanity at large. His responsibility for the fate of humanity could be reduced simply to his responsibility for the fate of science.

This limitation of the reflection on one's own activity to epistemological and methodological issues, and the identification of the methodological rules of science with the ethics of the scientist, led scientists to believe that science is only a disinterested search for truth rendered possible by a single methodology, and at the same time, to demand that the powerful and the wealthy support scientific research for the sake of its practical applications. This allowed them to treat science both as a system of ideas and as a means of achieving wealth, power, and control over the environment. There was no inconsistency in holding both these ideas so long as the progress of knowledge, as well as the technical progress linked to it, were seen as an unmixed and unquestionable good, or so long as the modern ideal of science linking the cognitive and the technical functions of knowledge appeared unproblematic. The realization of this ideal could then appear as the coming of the "kingdom of reason," the dream of the Enlightenment philosophers. If human irrationality, resulting—according to Pasteur—in "barbarism, fanaticism and destruction," is simply a result of the imperfections of human thinking, then the popularization of the scientific attitude and its extension to all areas of life can indeed provide a means by which all human conflict can be resolved.

For a scientist accepting these assumptions there can be no moral conflict between the goals or consequences of the development of science and the means leading to this development. In its practical dimension, science appears to him as a means of rationalizing social life, while in its intellectual aspects it appears as the rational method for attaining truth, and thus as a necessary condition for realizing these practical goals. If there are no conflicts between the rationality of ends to which the development of science leads in view of cultural values, and the rationality of the means by which these goals are to be realized, then science must be treated as the embodiment of human rationality.

Accordingly, doubts in this matter did not appear, at least not among scientists, until both the autonomy of science as a social institution and the autonomy of the knowing subject were called into question; until the connection was broken between the moral value of gaining new knowledge and the value of its applications in practice; and until it appeared that the social consequences of the use of the rational methods of investigation could not always be considered rational from the point of view of the very cultural values whose realization science was supposed to promote. Once these links were severed, the issue of the effectiveness of the methods leading to the achievement of acknowledged goals became separated from the issue of the unconditional acceptance of these goals. As a result, the problem of the rationality of science became two-dimensional, and each dimension deserves to be treated separately: first, there is the problem of the rationality of the scientific method based on the assumption of the autonomy of the knowing subject; and secondly, there is the problem of the rationality of the consequences of applying this method given the social status of science within the global social structure. For an adherent of scientism these two problems always merge into one—the first one. Because of scientism's unquestioning adherence to the modern ideal of science, the second problem has from this point of view been solved *a priori* and does not usually need to be reflected upon. An adherent of scientism is interested in the criteria of scientific progress, and not in the issue of whether this progress is good or bad for humanity, since this issue has already been decided for him. But this very separation of these two problems led to a crisis in the scientistic ideology.

Thus, when in the course of the twentieth century the autonomy of science as a social institution appeared more and more fictitious, the scientistic position began to change into a completely different ideology. Scientism came to imply the acceptance of external rather than internal jurisdiction over the choice of the goals of scientific activity. It came to imply the acceptance of a situation in which the scientist is transformed from an independent thinker into an expert giving advise about how to achieve goals that were set without his participation. In this new situation, the scientist claiming that "my ethics is my methodology" is renouncing his moral responsibility for the consequences of his activities. The adage "do well what you do" becomes suspect whenever the consequences of this activity become morally ambivalent.

Finally, we should note that the scientistic position, as I have presented it here, has been criticized from two quite different points of view.

First, it has been criticized for universalizing the acceptance of those values with which—it is believed—scientific work is inextricably linked, and which—in the view of scientism's critics—do not deserve to be universalized in this manner. This type of criticism, generally speaking, shares the scientistic image of science; but while scientism for this very reason values the scientific attitude positively, its critics evaluate it negatively. From this point of view, scientism would always be attacked by the adherents of those ideologies which postulate the organization of social life on the basis of the acceptance of certain claims and values as inviolable and not subject to intellectual critique: ideologies limiting to a greater or lesser degree the right of individuals to freedom of opinion, to criticism, or to the placement of universal interests over the particular interests of nations, classes or groups. The criticism of scientism becomes in this case not so much a rejection of a given view about science which the adherents of scientism accept, as a refusal to accept values which scientism located in science and which it linked unconditionally with science. As a result, scientism can become a convenient label for political or ideological opponents, and in this manner can be used to mean pretty much anything.[21]

Secondly, the critique of scientism can result from a conviction that although the values defended by scientism are worth defending and universalizing, and not just for the benefit of science alone, it is still a mistake to believe that the development of science leads automatically, always and everywhere, to the achievement of this goal, and that science is therefore an unalloyed moral good, or even that it is morally neutral. If—as such a critique suggests, and as I have tried to emphasize—science, like all human activity, is morally ambivalent, then the following of methodological rules cannot exhaust the moral responsibilities of the scientist; his responsibility cannot be limited only to his responsibility for the development of science, nor his self-knowledge only to methodological considerations in which the consequences of his activity can be regarded as morally unproblematic.

The values defended by scientism, the realization of which is linked directly with the development of science, can be defended even if one does not share the scientistic view of the development of science. They can be defended simply on the grounds that they are worth defending—although they cannot be justified absolutely either by science or in any other way.

The processes discussed above: the change in the social status of science as an institution, the undermining of the idea of an autonomous knowing subject, and the transformations in the self-knowledge of scientists which followed—though with some delay—from these changes, all contributed to the realization that the idea of science as the embodiment of human rationality is not unproblematic.

CHAPTER VI

ESCAPE TO WORLD THREE

1.

In the foregoing analysis I have tried to explain how it happened that the previously unproblematic conviction that science and its development are rational has become in our time a subject of a lively philosophical controversy. Here I will discuss one of the fundamental issues in this controversy, namely the idea that the rationality of science and its development is guaranteed by the rationality of the scientific method, since this method also constitutes the logic of the development of scientific knowledge.

In other words, we will consider here the position which attempts to defend this thesis, avoiding epistemological relativism while taking into account the problems involved in retaining the idea of an autonomous, rational knowing subject. There are also arguments supporting this thesis which proceed by reducing the rationality of the knowing subject to a technical rationality (that is, to effective action) and thereby reducing the rationality of science to its instrumental role. We might argue that while the first position constitutes an attempt to defend nineteenth-century scientism without taking into account the changed social status of science as an institution, the second view attempts to present the rationality of science in such a way that science is shown to fit this new status, to remain consistent with it, and even to sanction it.

To begin with, let us examine certain notions about the concept of rationality in general.

As long as the progress of knowledge and technological progress were treated as an unquestionable good, as long as the modern ideal of science uniting the cognitive and the technological functions of knowledge had the highest moral sanction, the ideal of science appeared unproblematic. Because this ideal was based on the concept of an autonomous knowing subject who, despite his individuality, was deemed naturally able to arrive at universally valid laws thanks to his inborn characteristics, there was no conflict, nor could there be any, between the evaluation of the consequences of the development of science and the evaluation of the methods leading to this development. The rationality of the subject—his assumed cognitive autonomy—sanctioned at one and the same time both the rationality of the goals and the rationality of the methods of science.

It can be said that although modern science distinguished programmatically between the world of facts and the world of values, in itself it pretended to embody both the true and the good. The realization of the ideal of science was to herald the coming of the kingdom of reason and goodness. This view expressed the idea that science is rational, which was philosophically grounded in the notion of the autonomous, naturally rational subject.

At the same time, we should note that although the concept of rationality was understood as an epistemological category, and was used to identify those cognitive procedures understood as inherently trustworthy methods of finding out the truth, in fact the concept of rationality functioned as a normative category used to identify cognitive procedures which were functional from the point of view of attempting to realize the accepted ideal of science. Since this ideal itself was treated as universally valid and unproblematic, *everything which appeared as a rational means of achieving this ideal was by the same token considered universally, always and everywhere, rational.* In this manner a normative and relative concept of rationality could appear to be a descriptive and non-relative category, and the difference between "rationality" and "irrationality" could seem a fact of nature—always and everywhere the same.

In a word, it was possible to claim that the rational method is as different from an irrational one as an owl is different from a donkey, and that *it is the task of epistemology or methodology to discover the nature of this difference*. In any case, it was believed that this difference was given once and for all, objectively, independently of what philosophers correctly or incorrectly believed this difference to be, just as in the classical theory of truth, where the truth of statements depends only on their content and not on our methods of evaluating them. Rationality could be considered an objective and supra-historical characteristic of specific cognitive procedures, a characteristic which they possessed immanently, and not from the point of view of the accepted ideal of science. It was believed that the claim that a given procedure is rational is a (true or false) judgment of the same nature as the claim that a given object is green or solid.

From this point of view, for example, if an explanation of physical phenomena by final causes is considered irrational today, then it was also irrational at the time of Aristotle. On the basis of such an understanding of the concept of rationality, the arguments against Copernicanism could never have been rational, even when they appealed to directly witnessed experiences, such as that if the earth were turning there would be a constant strong wind in the direction opposite to its movement, that a stone thrown from a height could not land directly underneath the place from which it was dropped, that wine would splash from a pitcher when poured, and that houses and fortresses would turn quickly into ruins. In a word, the victory of the Copernican theory was treated and presented as a triumph of the application of the rational method of investigation, not as the result of a revision of the accepted criteria of rationality—such as, for example, the highest authority of the direct data of the senses.

On the basis of this view it was impossible to claim that Copernicus was unable to defend his position with rational arguments,[1] since rational arguments tended to support his opponents. It is a fact that the Copernican theory could be defended by rational arguments after Galileo and Newton, since the scientific revolution which began in the sixteenth century changed the standards of rationality in science. But to raise this fact against the statement that Copernicus himself was unable to defend his theory with rational arguments is to invoke precisely such a supra-historical understanding of the concept of rationality.[2]

In short, I believe that the unproblematic acceptance of the prevailing ideal of science meant that the category of rationality, which was basically relative and normative, and historically contingent, was treated as if it were descriptive and supra-historical. If one does not maintain a certain distance from the accepted ideal of science (whatever it is a model of), then it comes to be seen as natural and indisputable and as the only possible one. It is then difficult to notice that the evaluations which this ideal imposes are indeed relative to this ideal. The understanding of the concept of rationality is not unique in this respect. A lack of distance from one's own culture leads as a rule to the belief that its conventions are natural, and that everything that does not conform to them is a violation of the natural order. Such is always the result of elevating the particular to the dignity of the universal.

Only when we become aware of the problematic nature of the accepted ideal—and we have seen how this came about—can we become aware (which does not mean that we are indeed always aware) that the designation of cognitive procedures as "rational" or "irrational" does not have the character of claims about natural facts. Only then do we notice that the statement that a given procedure is rational or irrational is not a descriptive judgment, referring to some immanent characteristics of these procedures, but a judgment evaluating them in terms of the prevailing ideal. Thus we might notice that the arguments against Copernicus, which appear irrational to us today, were completely rational with respect to the ideal of knowledge accepted by those who voiced them at the time. Only when the canons of common sense cease to appear natural can we understand that "if one does not sin against common sense," as Einstein said, "it is impossible to achieve anything." At the same time, we can become aware that *epistemology or philosophy of science does not discover the difference between rationality and irrationality, but rather constitutes this difference conventionally* (with respect to the accepted ideal), and decides which cognitive procedures will be deemed rational in terms of the realization of this ideal. If we fail to notice this circumstance, we are committing a "naturalistic fallacy," analogous to the fallacy described by G. E. Moore, which consists in assuming that value-predicates denote certain intrinsic characteristics of objects.

Thus the fact that a certain concept of rationality is considered obvious, universally valid, or even the only possible one, means only that the ideal of science which imposes this concept is regarded as unproblematic and that we are unaware of its historicity. Such a view of rationality, independently of whether it is held in good or bad faith, only conceals the relative character of this concept but does not change it. (When I refer to bad faith here I mean only to say that it is one thing to accept a given ideal in a context where it is generally considered unproblematic, and quite another to consider it unproblematic in a context where it is already seen to be controversial. In the second instance, it is not the defense of this ideal which is in bad faith but rather its uncritical acceptance.)

The difference between understanding the concept of rationality as a descriptive and supra-historical category, or as a historical and relative one, plays an essential role in contemporary discussions about the rationality of science and its development. I believe that we can formulate this more strongly: this controversy is as much a controversy about the mechanisms of scientific development as it is about the rationality of human cognitive activities. From this point of view it is not just a problem specific to philosophy of science, but a more general philosophical problem. Although I do not share the views of Imre Lakatos with respect to the substance of this dispute, I believe that he was correct when he argued that "The clash between Popper and Kuhn is not about a mere technical point in epistemology. It concerns our central intellectual values [...]"[3]

The fact that this dispute is still so lively in the philosophy of science testifies to the state of affairs which Krzysztof Pomian appropriately termed "la malaise de la science."[4] This situation signals a growing awareness that, on the one hand, the legitimation of the goals to which scientific reason is addressed escapes the jurisdiction of this reason, while on the other, evaluation of the rationality of the methods used is relativized to the acceptance of these goals. In this situation, not to notice the relativity of this evaluation, to treat it as unconditional and supra-historical, amounts to consenting to the abandonment of science's right to decide on its own goals, and the cession of these rights to someone else. Rationality is thereby reduced to effectiveness in realizing these goals. This is what I had in mind when I argued in the previous chapter that today scientism is no longer a defense of the autonomy of science, but a manifestation of a technocratic ideology: the technocratic order justifies the goals, while science is left to see to it that they are efficiently achieved.

2.

The question of what it means to say that science and its development are rational arises as soon as the cognitive autonomy of the subject is called into question. In order to address this question, we need first to make it more precise.

First of all, it is easy to see that the thesis that the rational character of the scientific method guaranteed the rational character of science (since this method constitutes at the same time the logic of its development) can easily turn into a tautology. So, for example, if we say that the philosophy (or methodology) of science formulates the rules of rational cognition, and that these rules constitute the demarcation criteria of science, then the rational character of scientific development is guaranteed by definition. Whatever fails to conform to the accepted standard of rationality (regardless of the nature of this standard) will simply not be considered scientific; and conversely, whatever cannot meet the criteria of demarcation and fails to agree with the accepted notion of science will not be treated as rational. Roughly speaking, this was Popper's position when he wrote *The Logic of Scientific Discovery* and prepared the expanded English edition of this work. His view was obviously supported by a normative conviction that modern science is the embodiment of human rationality.

This position, although it is logically unassailable in the sense that every terminological convention is unassailable (and Popper is quite aware of its conventional character), does encounter the following difficulty: it might turn out that the cognitive procedures used by scientists do not correspond—or what is worse, cannot correspond—to this conceptualization of the notion of rationality. They cannot correspond to it because the knowing subject is not autonomous. Apart from various epistemological ideas, this is precisely what has been shown by contemporary studies in the history of science which have revealed both the changeability of the scientific method and the extra-methodological conditioning of scientific development.

If, however, the difficulty were simply one of a basic divergence between the model of the rational development of science and its actual history, the position discussed above could still be defended as a normative conception, specifying not how science is actually pursued but how it should be pursued in order to be fully rational. But when it is claimed that not only is science not pursued in this fashion, but that it cannot be so pursued, then the normative interpretation of the model is threatened as well. It makes no sense to postulate norms which cannot be followed.

If the knowing subject cannot be completely rational—even when it is pursuing scientific research—then how can we defend the thesis that the rational method of science constitutes at the same time the logic of its development?

This can be accomplished, first of all, if one claims that the model of scientific development to be defended does not apply to the cognitive activities of a subject, but only to the objectified products of human activity, so that the concept of the subject is eliminated from the epistemology. By eliminating the subject, we eliminate also the possibility that any extra-rational, genetic, historical or social conditions of the subject could influence the development of scientific knowledge. This is the position defended by Popper in his theory of the three worlds and in his "epistemology without a knowing subject."[5]

It is easy to see that from this perspective the rationality of scientific development is again guaranteed by definition, since the elimination of the knowing subject from the area of concern of epistemology—its removal to world three—guarantees by the same token that no extra-rational factors could have any influence on the thus reconstructed process of scientific development.

The second possibility is to construct an "idealization" of the concept of rationality according to which all cognitive activity appears rational "at the limit." In order to do this, it is enough to accept as rational any activity which, on the basis of the knowledge possessed by the subject and in view of the goals he accepts, leads to the realization of these goals. There is no doubt that in this case the possibility of the rational reconstruction of the development of science is guaranteed by definition with the assumption that all human action (including cognitive action) is rational "at the limit."

In the first case, the idea of the rational development of science is saved by the elimination of the knowing subject, which cannot be described as autonomous and independent of extra-rational factors when engaging in cognitive activity; in the second case, the idea of rationality is saved by the concept of rationality itself, which at the limit makes all human cognitive activity rational, since all such activity is goal-oriented.

Considering these interpretations of the thesis of the rational character of scientific development and the possibility of its logical reconstruction by means of a given methodology, I must note that the meaning of the thesis being discussed changes if the term "rational method" is used to designate those cognitive procedures which are rational as such, that is, those whose rationality is not relativized to the accepted ideal of science, or to models of rationality understood in a relative manner.

In the first case, the question of why and how change affects these models of rationality in science is not and cannot be raised; and the process of scientific development is, from this point of view, seen to be continuous. Revolutions in science consist in the replacement of one theory by another, but these changes always take place as a result of the application of the same rules by which choices are made from among competing theories.

In the second case, the situation is more complicated. One could claim that the notion of the rational method of investigation is relativized to the accepted ideal of science, and that in this sense, it has a normative and evaluative character; and one could claim at the same time that this ideal is universally valid and historically constant, so that whatever does not conform to this ideal does not, by the same token, deserve to be called science. Although the naturalistic fallacy can thus be avoided, nevertheless one is still claiming that the development of science takes place (or at least can be reconstructed) as a process which always proceeds according to the same rules for discarding old and accepting new theories. This is how I understand Popper's position in both periods of his work. Submitting theories to strict attempts at falsification is considered by Popper as a rational and normative rule in view of the goals of science. Since these goals are considered unchanging, the norm has a universally valid character applying to all cognitive activity. Nevertheless, it is based not on the concept of the knowing subject but on a supra-historical idea of science.

The view that the criteria of rationality are relative can, however, be understood in terms more radical than those of Popper. It might be taken to mean that there are no universally valid (even normative) models of rational cognitive activity: that the development of knowledge cannot be treated as a continuous process, and that its rational reconstruction, even if only as an approximation to reality, is impossible.

3.

Today when we read Popper's *Logic of Scientific Discovery*, a book which first appeared more than half a century ago, we interpret it within a specific context of which it is good to be aware. It is impossible today to avoid having two frames of reference: the state of the philosophy of science half a century ago and its state today. Fifty years is long enough to make these frames of reference distinct, but probably not long enough for them to have changed completely. In short, both as a result of changes in the problematics of philosophy of science and as a result of the later work of its author, Popper's book is implicated today in polemics different from those it generated when it was first published. It is *already* impossible to read it as a contemporary book, and it is *not yet* possible to treat it as a venerable subject of study for a historian of philosophy. We read it today differently from the way we read, for example, Ernst Mach's *Erkenntnis und Irrtum*, and again differently from Jürgen Habermas's *Erkenntnis und Interesse*.

I chose these two titles not by accident. Their juxtaposition illustrates symbolically two types of philosophical reflection on science: one which treats science as an autonomous set of ideas developing according to its own autonomous "logic," and another which sees it not only as a system of ideas but also as a social institution co-determining the logic of its development.

The Logic of Scientific Discovery fits fully and programmatically into the first of these two types of philosophical reflection on science, and all of Popper's later work is a defense of the correctness of this approach, and of the view that the development of knowledge can and should be studied as an autonomous process, free from all extra-logical conditioning, and in this sense rational. Popper shared this point of view with his opponents from the 1930s. Years later, in his scientific autobiography, Popper himself wrote:

> But although especially Carnap's repeated demand for 'justification' was (and still is) to my mind a serious mistake, such a matter is almost insignificant in this context. For Carnap pleads here for rationality, for greater intellectual responsibility [...] It is this general attitude, the attitude of enlightenment, and in this critical view of philosophy—of what philosophy unfortunately is and of what it ought to be—that I still feel very much at one with the Vienna Circle and with its spiritual father, Bertrand Russell. This explains perhaps why I was sometimes thought by members of the Circle, such as Carnap, to be one of them, and to overstress my differences with them.[6]

The main subject of the polemics at that time (which I do not propose to examine here) can be presented in short as follows: according to the views of logical empiricism, which was criticized in *The Logic of*

Scientific Discovery, scientific knowledge should be based on purely empirical observation statements which are dictated by experience (and by linguistic rules) and therefore require no further justification. These statements alone constitute a frame of reference for the justification (verification) of all other scientific statements. Statements which are not reducible to observation statements, or (possibly) translatable into such statements, cannot be verified, and thus are devoid of all empirical meaning. They tell us nothing about the world, have no scientific status, and should thus be eliminated from science. The task of the theory of knowledge is, first of all, to reconstruct the observational language in which observation statements can be formulated, and secondly, to specify the logical relations obtaining between observational and non-observational statements, so that the first can inductively justify (verify, validate) the second. The achievement of this task would reveal the rational structure of all scientific knowledge, and would supply a model to which all intellectual achievements would have to conform.

In contrast to this programme, Popper's falsificationism negated the possibility of designating a purely observational language. It argued that all scientific claims have a theoretical character, and that there are no unquestionable empirical sentences. Basic statements, which can constitute the set of future potential falsifiers of scientific theories, are accepted by convention and are conventionally defined. The rationality of science is based not on rules of inductive confirmation, but on rules for the deductive elimination of false statements, that is, on principles of scientific criticism. Statements which are not subject to falsification, although they are not nonsensical, have a metaphysical character and should be eliminated from science (later Popper basically abandoned this claim in favor of the idea of metaphysical research programmes).[7] Theory of science has the task of formulating rules for the elimination of false theories and for their re-placement by better ones. The task of epistemology is thus to reveal the rational mechanism of the development of scientific knowledge.

The polemics between adherents of these positions took place within the framework of a broader consensus, as I have indicated above. The acknowledgement that the development of science and of its methods is conditioned not only by the unchangeable rules of the logic of scientific discovery, but also by psychological, historical or sociological factors, did not and does not fit into the perspective of Popper's falsificationism any more than it fits into logical empiricism. The acceptance of this view would for both of these perspectives be tantamount to a denial of the thesis of the rationality of science, and would deprive science of its claim to special status as a model of rational activity for the rest of human culture.

As is well known, this position found expression in the conviction that the task of philosophy of science was to study only the context of justification, and to exclude all questions concerning the so-called context of discovery. This position programmatically separated epistemology (theory of knowledge) from the sociology and history of science, and this was the case for both the variant in which the task at hand was seen as the reconstruction of the structure of an already existing knowledge, and for the variant in which the main concern was the reconstruction of the mechanism of its development. In this latter, properly Popperian version, logical reconstruction was to be a historiographic programme, a basis for the writing of the history of science.[8] This belief shared by both perspectives is known as the thesis of the primacy of logical over genetic questions.

Popper has defended this position continuously ever since he first published *The Logic of Scientific Discovery*, and he regards it as equivalent to the elimination of psychologism and subjectivism from the theory of science.

In a section of *The Logic of Scientific Discovery* entitled "Elimination of Psychologism," we read:

> Accordingly I shall distinguish sharply between the process of conceiving a new idea, and the methods and results of examining it logically. As to the task of the logic of knowledge—in contradistinction to the psychology of knowledge—I shall proceed on the assumption that it consists solely in investigating the methods employed in those systematic tests to which every new idea must be subjected if it is to be seriously entertained.[9]

And, in the paper on "Epistemology Without a Knowing Subject," he wrote:

> My first thesis can be put by saying that in the present problem situation in philosophy, few things are as important as the awareness of the distinction between the two categories of problems—production problems on the one hand and problems connected with the produced structures themselves on the other. My second thesis is that we should realize that the second category of problems, those concerned with the products in themselves, is in almost every respect more important than the first category, the problems of production. My third thesis is that the problems of the second category are basic for understanding the production problems: contrary to first impressions, we can learn more about production behaviour by studying the products themselves than we can learn about the products by studying production behaviour. This third thesis

may be described as an anti-behaviouristic and anti-psychologistic thesis.[10]

I have no intention of returning to a polemic with this position. In my opinion the pertinent distinction drawn between genetic and logical questions does not imply the primacy of one over the other, nor does it imply that the philosophy of science is obliged to limit itself to the second type of questions. I have argued elsewhere that if it were to do so, it could not serve as the basis for a reconstruction of the development of science, since it would then programmatically avoid the issue of the variability of methodological rules. And in order to explain this variability it is necessary to go beyond the context of justification. Since this position is sometimes understood as a suggestion that a synthesis of the "logical" and the "genetic" approaches is possible, I will add only that I regard this as a misunderstanding. Such a synthesis would be possible only if questions about the contents of consciousness could be reduced to questions about the conditioning of cognitive activities, which appears nonsensical. Such nonsense is, however, not a consequence of the thesis that both a logical and a genetic analysis are necessary in order to explain the mechanisms of scientific development.

Here I am interested in a different issue, namely how the thesis about the rationality of the process of scientific development and about the possibility of its logical reconstruction should be understood from a perspective according to which logical questions are prior to all others.

Although the general sense of the two statements of Popper which we have cited is similar, since both of them stress the primacy of logical questions over genetic ones, nevertheless, from the point of view of our concerns here, these two statements are not equivalent. The first clearly states that logic has the task of investigating the methods used for systematic testing or for justifying new theoretical ideas. In other words, it has the task of investigating what scientists do (or should do) when they seek to justify their claims. In contrast, the second statement claims that problems of the logic of the development of knowledge concern "existing structures," that is—as I understand from the context in which this statement is made—the relationships between these structures.

In the first instance, the object of the theory of the development of knowledge is understood as a set of rules governing research activity; in the second, as a set of relations obtaining among existing structures. In the first case, the theory is a blueprint (however idealized) for real cognitive activities; in the second, it says nothing about such activities. Accordingly, in the first case, the rationality of scientific development is grounded in the rationality of the procedures employed to accept or reject theories, that is, in the rationality of possible human cognitive behavior; while in the second, rationality is embedded in the logical relations among completed theoretical structures—which obviously is not quite the same thing.

The idea of the rational reconstruction of the development of knowledge changes analogically. In the first case, in the words of Popper himself, it is a methodological analysis of the "corresponding thought-processes" by means of which "the scientist critically judges, alters, or rejects his own inspiration." Popper adds that "this reconstruction would not describe these processes as they actually happen; it can give only a logical skeleton of the procedure of testing."[11]

In the second case, the reconstruction does not apply to any thought processes, neither as they actually occur nor to their logical skeleton. It is rather, as Lakatos says, a "reconstruction of the internal history of science," that is, of history presented according to the logical relations which (according to the accepted criteria of rationality) should obtain, or possibly do obtain, among existing structures. This "internal history" encompasses completely the rational aspect of the development of knowledge, while the "external history" of deviations from this rational model caused by psychological, historical, or sociological—in other words, irrational—factors, "is irrelevant for the understanding of science."[12]

When I first read *The Logic of Scientific Discovery* in the 1950s, before all the controversies about the rationality of scientific development had really started, it did not even occur to me to treat the logic of discovery proposed by the author as anything other than a model of the actual research activities of subjects, and I do not know of anyone who interpreted it otherwise. I believe that this shift in Popper's views has been brought about by the necessity of abandoning the previously accepted idea of the knowing subject.[13] The new idea defends the rationality of the development of science without getting involved in the issue of the rationality of the subject: the question of the extent to which the subject can be independent of all extra-logical conditioning and is able to formulate here and now statements which are valid always and everywhere. In order to see more precisely the character of this change, however, it is necessary to focus for a moment on Popper's more recent work.

4.

According to the author of *Objective Knowledge*, we should distinguish among three worlds. The first is the world of material objects, the second that of subjective psychological activities and experiences (impressions, feelings, convictions, etc.), and the third is the world of the products of these psychological activities.

> World 3 objects are of our own making, although they are not always the result of planned production by individual men. Many World 3 objects exist in the form of material bodies, and belong in a sense to both World 1 and World 3. Examples are sculptures, painting, and books, whether devoted to a scientific subject or to literature. A book is a physical object, and it therefore belongs to World 1; but what makes it a significant production of the human mind is its *content* [...] And this content belongs to World 3.[14]

Although they are human products, the objects of the third world are independent of man: they "live their own life" and evolve. Existing theories give birth to new problems which were not intended or even expected by their creators, which are no longer our direct products and which we must discover rather than make up in world three. They belong to this world although no one has invented them or thought about them yet, although they are not a part of anyone's consciousness. And in this sense the third world transcends its creators; it is at the same time a super-human world.

> Although man-made, the third world (as I understand this term) is super-human in that its contents are virtual rather than actual objects of thought, and in the sense that only a finite number of the infinity of virtual objects can ever become actual objects of thought.[15]

> Its [the third world's] action upon us has become more important for our growth, and even for its own growth, than our creative action upon it.[16]

I will not discuss in detail the controversies surrounding Popper's interpretation of the ontological status of the third world. Despite some important differences, this interpretation is remarkably similar to Plato's notion of the world of ideas, to Hegel's objective Spirit, or to Frege's pure meanings:

> Bolzano was, I think, doubtful about the ontological status of his statements in themselves, and Frege, it seems, was an idealist, or very nearly so. I too was, like Bolzano, doubtful for a long time, and I did not publish anything about world 3 until I arrived at the conclusion that its inmates were real; indeed, more or less as real as physical tables and chairs.[17]

The reality of this world is finally determined, according to Popper, by the fact that it acts causally on the second world, and through it on the first world as well. In order to grant this—that the products of human intellectual activity are not just epiphenomena, and that the world of culture affects us and through us affects physical reality as well—it is not necessary first to assume the real existence of the world of concepts or problems. Nor need we waste time in showing that this idea entangles Popper in all the known problems of interactionism, which tries to use causal categories to explain the effects of the mental world on the physical world. (Popper's last book, written together with the neurophysiologist John Eccles, is an attempt to solve this problem.)

I am interested here exclusively in the statement that the theory of the development of science is to be the theory of the development of the third world, together with the idea of the knowing subject implied in this claim.

The knowledge of a subject, as Popper now claims, can never be free of various irrational and extra-rational constraints. This is why philosophers, even the best ones like Hume and Russell, concentrating traditionally on the investigations of the second world, were finally unable to avoid subjectivism and irrationalism. As Popper puts it: "we use objective knowledge in the formation of our personal subjective beliefs; and although personal subjective beliefs can always be described as 'irrational' in some sense, this use of objective knowledge shows that there need not be any Humean conflict here with rationality."[18]

This means, I think, that the rational character of the development of science can be defended only at the cost of the complete elimination of the subject from epistemology. The logic of scientific discovery, which was to be a logical scheme of the course of actual thought processes, is by the same token transformed into something like the immanent logic of the development of the third world, which is autonomous with respect to the subject. A confirmation of this interpretation can be found in Popper's explanation of the difference between his idea of the third world and its evolution and Hegel's idea of the objective Spirit. "In spite of a certain superficial similarity between Hegel's dialectic and my evolutionary schema [...] there is a fundamental difference," as Popper claims in *Objective Knowledge*.[19] There can be no doubt that his schema is indeed different; but it is a logic of the development of the same universe with which Hegel's dialectics was concerned: the universe of supra-individual and supra-human beings, whose existence was not accepted by the author of *The Open Society* and *The Poverty of Historicism*.

What, then, is the nature of the relationship between events taking place in the second and in the third world: between the course of human cognitive activities, which are always affected by the influence of certain extra-rational factors, and the fully rational development of the autonomous third world?

The idea that the third world is a product of man and has feedback effects on the subject seems clear and obvious, if it is understood to mean that people's participation in the world of culture and its objectified products conditions their thoughts, their convictions and beliefs; in other words, if it is understood to mean that the process of the creation and transmission of knowledge is socially mediated. What is special to the position discussed here, however, is the thesis that social mediation applies only to the universe of subjective knowledge—the second world, which is not the object of concern for epistemology—and not to the third world. Its logic of development is to be understood as independent of its genesis. This thesis allows Popper to maintain that the development of the third world can be fully rational despite the fact that knowing subjects are not fully rational, since their subjective knowledge is always affected by extra-rational factors. It is not difficult to notice that this thesis is simply a different formulation of the conviction that logical and genetic problems are independent of one another, and that logical problems have priority over genetic ones. The elimination of the knowing subject is justified by this thesis, which guarantees the rational development of scientific knowledge as an object of the third world.

The influence of the second world on the third world is limited to the "supply of raw material" and to the production of "mutations of ideas." This biological analogy is for Popper more than a literary trope. For him, the distinction between the second and third worlds actually marks the difference between the world of nature and the world of culture, separating the world of biology from the world of logic. Mechanisms of the development of the third world constitute a filter through which the "natural selection" of ideas takes place; this is how they demonstrate their fitness in the third world.

Thus Popper states:

> the growth of our knowledge is the result of a process closely resembling what Darwin called 'natural selection'; that is, *the natural selection of hypotheses*: our knowledge consists, at every moment, of those hypotheses which have shown their (comparative) fitness by surviving so far in their struggle for existence; a competitive struggle which eliminates those hypotheses which are unfit.[20]

The emergence of such a language would face us here again with a
highly improbable and possibly unique situation, perhaps as improbable
as life itself. But given this situation, the theory of the growth of
exosomatic knowledge through a conscious procedure of conjecture and
refutation follows "almost" logically: it becomes part of the situation as
well as part of Darwinism.[21]

The difference between the amoeba and Einstein is that, although both
make use of the method of trial and error elimination, the amoeba
dislikes to err while Einstein is intrigued by it [...][22]

Cultural development, we can say, extends genetic development by other
means, that is by means of the objects of the third world.[23]

The relationship between the second and the third world can be
understood roughly as follows: man as a biological being is equipped with
some inborn knowledge, which he—just like other animals—modifies by
the method of trial and error depending on the circumstances of his life.
This process taking place in the second world is not rational because it is
conditioned by various particularistic factors. It does, however, supply
material which then has to demonstrate its fitness in the third world of
objectified knowledge, a world which is a specifically human product in
the sense that only man creates such a world and is subject to its feedback
effects. The rational development of this world can be guaranteed by
scientific criticism, which constitutes a cultural extension of the biological
mechanisms of natural selection by trial and error. Objective knowledge
is thus an improved means of human adaptation.

The question of why such criticism is rational can be answered
simply: it is rational because it allows for the elimination of errors, false
beliefs and convictions without eliminating their carriers. Thus it leads
both to truth and to good. An animal that errs has to die. A human,
thanks to the fact that the logic of the third world influences or can influence his subjective behavior, does not have to die; he can correct his
opinions. The criterion of rationality is thus ultimately grounded in the
norm of the biological survival of the individual. From this perspective,
scientific criticism is both an objective mechanism of the development of
the third world and a norm of human behavior.

This is why I believe that the normative and the descriptive layers of
Popper's philosophy cannot be separated.

5.

Several additional issues need to be raised in connection with what has been said about the idea of an epistemology without a knowing subject.

First, *from this perspective the concept of rationality receives a biological legitimation*. Scientific criticism is seen as a cultural expression or extension of the mechanisms of biological development. This is a philosophically important thesis, and as far as Popper is concerned, it is new. Despite a number of differences, it is not difficult to notice here a similarity between Popper and Piaget's idea of genetic epistemology, or the ideas of Jacques Monod, Konrad Lorenz and Noam Chomsky. It seems that we encounter here a tendency, typical of our times, to ground human rationality in biology, as a means of opposing historical relativism by appealing to biological constants. (From a different perspective—for example that of Husserl—it could be said that historical relativism is being replaced here by biological relativism, since appealing to biological constants means that truth is made relative to species.) It is characteristic of all these ideas to treat knowledge—including scientific knowledge—as a tool facilitating the adaptation of the species to its environment. A full consideration of this thesis in its various formulations would, however, require analyses which go far beyond the limits of this work.

Secondly, while classical epistemology assumed the autonomy of a knowing subject who is by nature able to think scientifically in complete disregard of all particular (historical, individual or biological) conditioning, the conventionalist methodology which we know from the *Logic of Scientific Discovery* rejects this assumption in both its aprioristic and in its radical empiricist formulation. Conventionalism in all its variants was a critique of both of these currents of modern epistemology. It did not assume that the knowing subject was by nature a rational subject, but claimed that it could become one if it followed certain normative rules of cognitive behavior. Popper claimed at the same time that these rules allow us to reconstruct the "logical skeleton" of the actual thought-processes which result in the development of science.

Popper's subject was thus not rational by nature, but he could become rational as a result of applying definite normative rules of investigation. His actual research activities in science were treated as an area which was sufficiently close to the accepted model of rationality to make the program of logical reconstruction legitimate. At the same time, the methodological individualism of Popper dictated that the development of scientific knowledge be treated as the result of the individual cognitive activities of single subjects (just as history was to be treated as the outcome of the behavior of single individuals), and he strongly opposed the acceptance of any supra-individual entities. The entire Popperian critique of Plato, Hegel and Marx in *The Poverty of Historicism* and *The Open Society* was framed in these terms.

The passage from *The Logic of Scientific Discovery* to the theory of the development of the third world means both the *abandonment of methodological individualism* (or at least its substantial weakening, since the third world is a supra-human world), and the *abandonment of the earlier conception of the subject. The subject is now perceived as relative to his existential conditions, and his cognitive activities can no longer be seen as sufficiently close to the models of rational behavior* to be used as a basis for the reconstruction of the development of science. Individual thought processes supply only a subjective material which in its objectified form can evolve rationally in the supra-human world. It is as if the idealization of the knowing subject as a rational subject had been taken to its ultimate limits, that is, to the complete elimination of the subject as a topic of interest for epistemology and theory of science.

I believe that this change in Popper's views was caused, among other things, by the fact that contemporary studies of the development of science and of the socio-historical conditioning of this development no longer allowed him to defend the idea of the rational subject which he had previously upheld. More precisely: Popper had to accept the fact that this conception was not a good idealization of actual human cognition, and that the relations among the products of cognitive activities were not a sufficient basis for an account of the course of actual thought processes. *If this is so, then either it is necessary to give up the idea of the rational character of the development of science* (as it was understood until then) *or to claim that science is rational even if the cognitive activities of individual subjects are as a rule influenced by extra-rational factors. This is how I understand Popper's path to the third world.* As mentioned earlier, the process of the development of science reconstructed in the third world is by definition a rational process. No individual, biological or historical factors can have any influence on the processes taking place in this world. It is baffling how Popper manages to reconcile his "Darwinian" theory of the development of knowledge as a quasi-biological process with his insistence that this development can and should be accounted for without reference to any genetic explanations.

It seems to me that the source of these difficulties lies in the ahistorical treatment of the concept of rationality, and the simultaneous conviction that the development of scientific knowledge is the embodiment of rationality and indeed its model. It should be noted that the same ahistorical concept of rationality is used also by those who—when faced with the alternatives presented above—are prepared to follow a path different from that of Popper and conclude that the development of science is not a rational process, or—more carefully—that it is not fully rational. I will discuss this view further in the final chapter.

A third comment: *The recognition of scientific criticism as the mechanism of the development of science and as a norm of rationality clearly serves to guarantee the special status of science in the third world, that is, in the world of culture.* From this perspective scientific criticism plays the same role that the idea of the autonomous knowing subject played earlier. Knowledge developing according to the rules constitutes both the realization of the good and the elimination of falsehoods (thus approaching ever more closely to the truth). Both are possible because human thought does not belong to the second world. As Popper says:

> Our hope is thus that traditions, changing and developing under the influence of critical discussion and in response to the challenge of new problems, may replace much of what is usually called 'public opinion', and take over the functions which public opinion is supposed to fulfil.[24]

In other words, Popper's hope is that his ideal of science will become a model for all human culture and will promote its rationalization. Scientific criticism will become a basis for the settlement of all issues which previously were, or were supposed to be, decided by public opinion, which as a rule is not sufficiently critical and not sufficiently rational. Popper is in this sense an heir of scientism regardless of the distance between his views of the scientific method and nineteenth-century opinions on this subject among followers of scientism, and regardless of the sincerity of Popper's dislike of the term. (Popper's correction to Hayek's definition of scientism, which I mentioned above, is in this context quite revealing.)

And finally, a fourth remark: From Popper's perspective *scientific criticism plays not only the role of the mechanism of the development of the third world, but also the role of a moral norm.* There is no doubt that despite all the changes which Popper's philosophy has undergone over the years, it could always be read as expressing an unchanging ethical proposition. If I were to express this ethics in one concise formula, I would claim that both in his philosophy of science and in his social philosophy Popper proposes an ethics of a certain minimalism. As Popper says,

If I were to give a simple formula or recipe for distinguishing between what I consider to be admissible plans for social reform and inadmissible Utopian blueprints, I might say:

Work for the elimination of concrete evils rather than for the realization of abstract goods. Do not aim at establishing happiness by political means. Rather aim at the elimination of concrete miseries. [...]

But do not try to realize these aims indirectly by designing and working for a distant ideal of a society which is wholly good. [...] Do not allow your dreams of a beautiful world to lure you away from the claims of men who suffer here and now. [...] no generation must be sacrificed for the sake of future generations, for the sake of an ideal of happiness that may never be realized. In brief, it is my thesis that human misery is the most urgent problem of a rational public policy and that happiness is not such a problem. [...]

It is a fact [...] that it is not so very difficult to reach agreement by discussion on what are the most intolerable evils of our society, and on what are the most urgent social reforms. [...]

With ideal goods it is different. These we know only from our dreams and from the dreams of our poets and prophets. They cannot be discussed, only proclaimed from the housetops. They do not call for the rational attitude of the impartial judge, but for the emotional attitude of the impassioned preacher.[25]

The Popperian ethics of cognition based on the norm of scientific criticism could be formulated in a similar language. It would say: Act so as to eliminate specific error rather than to achieve absolute truth. Falsehood can be definitively proven, while with truth, even if we do possess it, we can never know that we really do. Thus make sure that your claims are susceptible to falsification, and right from the start specify the conditions under which you would be ready to abandon these claims. Do not let your dreams of the truth prevent some judgment from being rejected only because this judgment seems to you to correspond to the ideal you are searching for. No falsehood today can be tolerated in the name of a future truth which perhaps is unattainable. The elimination of falsehood is the most urgent problem of cognition; reaching the truth is not so urgent.

This is an ethics which I clearly find very appealing. My own objections and doubts, which I have expressed both here and elsewhere, concern not this ethical standpoint, but rather the descriptive claims about the actual development of science, which Popper believes to be governed by methodological rules which correspond to this ethics and legitimate it. The changes in Popper's philosophy discussed in this chapter can be interpreted as a defense of this legitimation which today is being questioned in a variety of ways. In other words, *I do not believe that it is possible to legitimate this attractive ethical posture with references to science, or that the development of science provides such an unchanging model of a rational attitude.*

But the fact that the development of science does not proceed in reality according to Popper's methodological rules by no means invalidates Popper's ethics. At present the acceptance of this ethics is probably far more important than a recognition of the image of scientific development to which it corresponds.

CHAPTER VII

ARE THERE SELECTION CRITERIA?

1.

Popper's supra-historical concept of a scientific reason engendering science in its own manner justified the programme for developing the philosophy of science (methodology) as a logical reconstruction of the development of science. Our critique of this conception is based on the idea that the criteria of rationality have neither a descriptive nor a supra-historical character, so that the logical reconstruction of the development of knowledge cannot give an account either of the course or of the mechanics of this process. Although this view was stated explicitly in the first chapter, it was not properly justified. More precisely: I presented historical arguments which—in my view—support this critique, but historical arguments cannot replace methodological analysis. They cannot do so if we do not want to move from one extreme to another: from treating the context of justification as the only legitimate object of philosophy of science to the treatment of historical processes as justifications.

The problem of the existence of criteria for choosing between competing scientific theories is one of the most controversial methodological issues, and its analysis exemplifies the contrast between the two opposing views of rationality mentioned above. Generally speaking, this problem can be formulated as the question of whether or not a methodology is able to explain the process by which one theory is replaced by another in terms of the application of definite, universally valid rules of research practice, rules which could be codified and presented as rational.

It is obvious that a negative answer to this question defeats the program of the logical reconstruction of the development of knowledge as a goal of methodology, since such an answer implies that in order to explain the passage from the old to the new theory it is necessary, at least in some instances, to appeal to extra-methodological factors, and thus to go beyond the context of justification.

The question has also been formulated differently, namely: can the accepted methodology be used to indicate the rational method for making a choice in every situation in which a choice between competing theories is required? Does the methodology imply an heuristic programme for such situations, one which every scientist should follow?

The second essential problem in this controversy is that of the mutual relations between successive theories in a given field: the so-called problem of the *correspondence of theories*. If it is the case, as some authors claim, that a relationship of semantic correspondence does not always obtain between successive theories, then it is clear once again that a logical reconstruction of this type of theoretical change is not always possible; and it would be pointless to insist on methodological rules demanding that new theories must always be constructed in such a manner that the requirement of semantic correspondence with their predecessors be met.[1] This issue, however, will not be addressed until the following chapter.

The discussion of both of these problems has to be prefaced with a remark. Although I do not believe that problems, and especially philosophical problems, can ever be solved definitively, nevertheless, without further justification, I dismiss here the position of radical empiricism. That is, I believe that no empirical fact can be recognized without being entangled in some prior theoretical views. The consequences of this decision, which I have justified elsewhere, are not the same for the analysis of the two problems I have mentioned.[2] The first problem—the issue of the existence of selection criteria—appears in methodology regardless of whether we accept or reject the position of radical empiricism. The problem of correspondence between theories, on the other hand, is not controversial within the framework of radical empiricism: this correspondence is guaranteed by the existence of an empirical basis independent of both of the successive theories, a basis with which both of them have to agree.

2.

The problem of criteria for choosing among competing theories can be formulated as the question of whether crucial experiments are possible in science. Are there experiments which could conclusively verify one of the competing theories, or conclusively falsify all the competing theories except one? Since, as is well known, both conclusive verification and conclusive falsification depend on the unquestionable nature of the empirical evidence to which one appeals for this purpose, then it is clear that it suffices to question the existence of such evidence in order to be able to reject the thesis of the possibility of conclusive verification or falsification. In referring to the unquestionable nature of the evidence I obviously mean epistemological and not technical incontestability. The issue is not whether it is possible to obtain evidence that is free of all experimental error (imperfection of the measuring apparatus, error in the reading of data, etc.), but whether unquestionable evidence could be obtained were it not for the ever present possibility of experimental error. The answer to this question is linked with the controversies in radical empiricism concerning the existence of purely empirical statements which are not laden with any theoretical assumptions. The existence of such statements is only a necessary, but not a sufficient condition for conclusive verification or falsification (whether or not they are conclusive will also depend on the logical relations between the ascertaining of facts and the acceptance or rejection of a theory); so I shall reformulate the question we are addressing here in such a way as to make it independent of the controversy about radical empiricism. The question can thus be formulated as follows: even if we disregard the possible epistemological fallibility of empirical statements, are these statements capable of the conclusive verification or falsification of a theory? And can they, by the same token, supply criteria for choosing between competing theories?

The thesis of the impossibility of conclusive verification, even given the assumption of unquestionable empirical evidence, is almost universally accepted today. This follows from the fact that induction is not a conclusive type of reasoning, and the verification of theories, even on the basis of indubitable empirical evidence, has to rely on induction. In essence verification takes place according to the logical scheme:

$$(H \rightarrow E) \cdot E$$

Since a theoretical statement (the hypothesis H) has an infinite number of empirical consequences E, no finite number of experiments ascertaining the actual occurrence of E constitutes a conclusive verification of H. In short, empirical statements, even if they were beyond doubt, could confirm H to some extent, but they could not verify it conclusively.[3]

These were among the reasons for the rejection of the postulate that only conclusively verifiable statements should be considered scientific, since according to this postulate all universal statements (laws and theories) should be denied this label and treated as unscientific, metaphysical, or devoid of empirical content as rules of reasoning.

Nevertheless, the logical scheme which I cited above seems to indicate that while conclusive verification is impossible, conclusive falsification is possible, since:

$$[(H \rightarrow E) \cdot (\sim E)] \rightarrow (\sim H)$$

If a testable universal statement implies the occurrence of E, and in fact we ascertain (~E), then—assuming the trustworthiness of this evidence—we must conclude that H is false. A false consequence cannot follow from a true statement. Here we have an instance of clear *logical* asymmetry between the confirmation and disconfirmation of deductively derived inferences, on the basis of which decisions are to be made about the truth of general statements. This much is also recognized in the common claim that a sum of examples does not add up to the truth of a general statement, while one counter-example is sufficient to refute such a statement.

For Popper, this logical asymmetry was the starting point for the formulation of the rules of the methodology of falsificationism. Thus, if the logical scheme cited above corresponds with actual attempts to test theories in science, crucial experiments would be possible, and it would be possible to formulate a rule requiring that in choosing among competing hypotheses we accept the one which survived (at least provisionally) the crucial experiment—provisionally, since we cannot exclude the possibility that in a confrontation with yet another theory, it may fail to survive the next experimental test. Thus, although the acceptance of a hypothesis which survived a crucial experiment is not conclusive, the elimination of its competitor is final; and the attempt to submit a theory to the most severe empirical tests would constitute the rationale of a scientific method to insure the rapid progress of knowledge by the elimination of falsehood. This method could at the same time serve as the basis for a historical reconstruction of the process of replacing old theories by new ones.

In his classical work *The Aim and Structure of Physical Theory*, Pierre Duhem questioned the opinion that the logical scheme described above (*modus tollens*) corresponds in fact with the procedure for the testing of hypotheses. As Duhem wrote: "the physicist can never subject an isolated hypothesis to experimental test, but only a whole group of hypotheses; when the experiment is in disagreement with his predictions, what he learns is that at least one of the hypotheses constituting this group is unacceptable and ought to be modified; but the experiment does not designate which one should be changed."[4]

According to Duhem's thesis, the procedure of testing follows the scheme:

$$(H \cdot A) \rightarrow E$$

In the case of a positive experiment we get:

$$[(H \cdot A) \rightarrow E)] \cdot E$$

which, however, does not imply the truth of the conjunction H·A, not only because this conjunction has infinitely many consequences, some of which might turn out to be false, but also because a positive result of an experiment can be achieved even when both parts of the conjunction are false: a conjunction of two false statements might imply a true statement. In the case of a negative result of an experiment we get:

$$[(H·A) \rightarrow E]·(\sim E)$$

which implies that at least one of the parts of the conjunction H·A is false, but does not specify which one.

If this is a scheme of testing (and today nobody seems to deny that the testing of an isolated hypothesis is impossible, and that there are always some background assumptions), then the conclusion must be that *crucial experiments are impossible.*

In a situation of two competing hypotheses H_1 and H_2 (tested together with the accompanying sets of assumptions, A_1 and A_2), when consequence E_1 is falsified and consequence E_2 is confirmed, we cannot conclude either that H_1 is false (because what might be false are the assumptions A_1), nor that H_2 survived a crucial experiment and should be retained. We can obtain the same experimental result when H_1 is true, if A_1, H_2 and A_2 are all false. By the same token, we are unable to formulate an unequivocal rule for choosing among competing hypotheses (theories), since we do not know whether we should replace H_1 with H_2 or rather reject H_2 and retain H_1, modifying the background assumptions A_1.

The logical asymmetry noted above does not guarantee an asymmetry in the procedures of verification and falsification; on the contrary, it turns out that falsification, as it actually takes place, is no more conclusive than verification.

Duhem's thesis was radicalized by Quine, who claims that on the basis of an experiment not only we do not know what has been falsified, but that moreover it is always possible to find background assumptions A_1' such that the conjunction $H_1·A_1'$ would imply ($\sim E$), that is, that it will imply the result of the experiment which was supposed to have falsified H_1. This means that hypothesis H_1 can always be protected from falsification by the introduction of a different background assumption A_1' in the place of A_1. Such a modification of assumptions used to be called *ad hoc.* While Duhem's thesis states only that we cannot determine whether what is false is the hypothesis or the background knowledge, Quine's thesis states additionally that a hypothesis threatened by the results of an experiment can always be saved from the verdict of the experiment by the introduction of appropriate modifications in the background knowledge (*ad hoc* hypotheses), and these modifications might even concern rules of reasoning.

When Popper formulated his thesis of the asymmetry of falsification and verification he was, of course, familiar with Duhem's thesis, even in the more radical formulation which I have ascribed here to Quine, but which in various versions was also formulated earlier by the conventionalists.[5] Popper's defense of the position that falsificationism supplies criteria for choosing between competing theories took various forms in different periods, and I will not discuss them here in historical order.[6] For my present purposes it is better to consider the arguments against the Duhem-Quine thesis in a certain logical order. These arguments can be listed—from the strongest to the weakest—as follows:

a) the Duhem-Quine thesis is false;

b) although the Duhem-Quine thesis is correct, it is not permissible to try to save scientific theories from falsification by the introduction of *ad hoc* hypotheses;

c) it is impossible to exclude from science all *ad hoc* hypothesis (modifications of background knowledge), but it is possible to formulate conditions which allow us to distinguish permissible modifications of background knowledge from those which are not permissible (the term *ad hoc* hypothesis would then apply only to the impermissible modifications); and

d) it is necessary to distinguish the methodological evaluation of theories from heuristic rules of selection, and from the problem of who is the addressee of these rules—the individual scientist or the scientific community. It can happen that although a methodological evaluation of theories does not allow for the formulation of any heuristic rules for the scientist, it does allow for the formulation of such rules for the scientific community as a whole.

As opposed to the first three arguments, the fourth argument does not appear at all in Popper's work, but only in the work of some of his students, who attempt in this manner to avoid the admission that the heuristic value of methodology is basically minimal since methodology is unable to supply us with any rules for selecting theories.

If none of the listed arguments turned out to be correct, then we would have to conclude that methodology can in no way codify rules for choosing among competing theories, that is, that it is incapable of telling us which choices are rational and which are not. In that case, we could not expect ever to be able to reconstruct the process of the development of scientific knowledge on the basis of methodological rules. Alternatively, we would have to admit that whatever choices a scientist makes, he is always behaving rationally; or to accept the fact that if there are no rational rules for choosing among theories, then the development of knowledge is not a rational process in this sense.

In the following sections of this chapter, I will try to show that there is no satisfactory defense of the view that unequivocal criteria for choosing among competing theories can be formulated.

3.

In discussing the arguments against Duhem's thesis it is important to keep in mind what this thesis does *not* assert. It is occasionally rejected on the basis of arguments which, even though ostensibly correct, are not in fact relevant to it.

First, Duhem's thesis does not deny either explicitly or implicitly that in the history of science some results of experiments have been treated as conclusive falsifications of certain theories—for example: geocentrism (the confirmation of stellar parallax), the theory of cosmic ether (the Michelson-Morley experiment), the theory of phlogiston (Lavoisier's experiments), and many others less spectacular. *The fact that in the history of science some experiments were treated as conclusive falsifications is not a valid argument against the thesis under discussion.*

Secondly, Duhem's thesis by no means asserts that whenever scientists groundlessly considered a given experiment as a conclusive falsification of a theory, they were making a mistake in their choice of theory. The fact that they chose correctly does not mean that the experiment was conclusive; and this is what is at stake in the controversy about Duhem's thesis. One can choose correctly even on the basis of an inconclusive experiment, just as knowing the truth and knowing that one knows it are two different things. *Thus, neither the fact that certain experiments were treated as crucial, nor the circumstance that sometimes on the basis of such experiments scientists made correct choices among competing theories, can be used as arguments for the possibility of crucial experiments in science.* In order to present a valid argument against Duhem's thesis it must be shown that these experiments were in fact conclusive, that is, that the choices based on them were *the only possible* ones, and that they were made in conditions of *certainty* as to the falsity of the rejected theories, rather than on the basis of subjective assumptions about this falsity.

That this was not the case can be shown by theoretical arguments which we will examine shortly. But an indication of the lack of validity of such historical arguments can also be found in the fact that certain experiments which at some point were considered decisive later had to be dubbed inconclusive, either because certain background assumptions were not taken into account or because certain possibilities for modifying the theory were not considered. Authors who reject Duhem's thesis by claiming that there are many examples of crucial experiments in science are generally reluctant to invoke those experiments which were groundlessly considered to have been crucial.

As an example we can use Pasteur's famous experiment supposedly falsifying the theory of spontaneous generation or abiosis. We know that in 1938 Oparin formulated, and in 1958 Urey extended a theory according to which life on earth originated through abiogenesis. It is not necessary to demonstrate that Pasteur's experiment was well known to both of these scientists. Without entering into the details of these experiments it is easy to notice that Pasteur's experiments could falsify abiogenesis only on the assumption of the correctness of his premise that if spontaneous generation were possible, then bacteria would appear in his samples during the time and under the conditions of his experiment. (This assumption was obviously accepted not only by Pasteur, but also by all those who accepted his experiments as conclusive.) It is enough, however, to assume that abiogenesis demands different initial conditions—the presence of methane, ammonia, water and hydrogen in the atmosphere, along with ultraviolet radiation which might lead to the formation of amino acids—and to admit that the process itself takes a very long time, in order to demonstrate that Pasteur's experiments cannot undermine the theory of abiogenesis. In any case, after these experiments there was no logical necessity to abandon the theory of spontaneous generation and to accept the principle that life comes only from life. One could just as well modify the theory of abiogenesis by introducing into it various additional *ad hoc* hypotheses. At the time of Pasteur, the hypothesis that ultraviolet radiation is a factor promoting abiogenesis would certainly have been considered *ad hoc*.[7]

Thus, Duhem's thesis has an epistemological and not a historical character. It asserts only that no experiment which in the history of science was, is, or will be taken as a conclusive falsification of some scientific theory, can ever in fact be conclusive. And if scientists have sometimes managed to make correct choices on the basis of such experiments, they were nevertheless mistaken if they thought that they were not risking error in making them; and they were mistaken not because their experiment was imprecise or improperly conducted. Only theoretical arguments can decide whether the belief that an experiment is conclusive is correct or not. The circumstance that this conviction was widely shared at the time of the experiment is not decisive. Thus, I cannot agree with Jan Such, who argues that a theory can be considered as definitively falsified when it is no longer defended by anybody, although he observes correctly that some theories in science have been *de facto* abandoned by everyone.[8]

It is also not a solution to the problem we are interested in here to claim, as Such does, that "the decisive situation is not so much a unique act of falsification of a given theory as it is a process of deepening contradictions between the theory on the one hand and other knowledge and experience on the other."[9] The problem with the possibility of conclusive falsification, however, is precisely that of knowing when the "deepening contradictions" are deep enough for a theory to be unconditionally rejected, and when they are not yet so deep as to force us to do so. The otherwise correct assertion that in their actual practice scientists do not reject theories on the basis of a single negative confrontation with the results of an experiment (especially if they have been considered well grounded until then) does not yet eliminate the question of why, before the occurrence of this last event in the process of "deepening contradictions," the theory was not yet conclusively falsified, and after the event it clearly was. When the rejection of a theory is seen not as a single experimental act but as a process, a series of experiments, the problem of the existence of the crucial experiment does not disappear, but rather centers on the last element in the series.[10] The question "How little hair does a person have to have to be considered bald?" cannot be answered with the otherwise correct "dialectical" argument that balding is a process.

Jan Such correctly defines the issue under discussion as the problem of the possibility of the definitive falsification of some fragment of theoretical knowledge, as a result of which we are certain that the fragment is false. But while claiming that there are decisive theoretic-experimental situations which make possible the definitive falsification of a scientific theory, he nevertheless writes, "we must recognize the fact that the crucial experiment is a typological and gradual, rather than a classificatory, concept and that none of the actually performed experiments can be considered crucial in the traditional (or more precisely, methodological) sense of this word, that is, as definitively decisive.[11] As for what Such calls "decisive situations"—those which constitute an outcome of the deepening of contradictions—they are "not created by more or less crucial experiments, but co-created by the more or less decisive theories which are built around them."[12]

But the issue is precisely whether they are indeed "definitive" or only "more or less" decisive. If we refer to the concept of a crucial experiment and of a decisive situation as being definitively decisive, but in the same breath speak of them as being more or less decisive, then the claim that these are typological concepts does not really help much. Such a typological concept refers both to what it is supposed to refer to (decisive situations) and to what that concept is being contrasted with (non-decisive situations or more or less decisive situations). When the theoretical conclusiveness of falsification is the subject of dispute, one cannot justify it by claiming that a very high level of certainty is the same as practical certainty, since it is not the practical but the theoretical certainty which is at issue. It might be very interesting of course to investigate the circumstances under which scientists have believed that they did possess such practical certainty or even thought they had theoretical certainty; but this is not an argument for the claim that Duhem's thesis is incorrect, or for Such's attempt to claim that the situation is *de facto* decisive.

4.

If one accepts that no theory or hypothesis is ever tested in isolation (and today there is almost total agreement that this is the case), then Duhem's thesis can be rejected only if we decide that all theoretical assumptions (all elements of background knowledge) are either certain or have no influence on the interpretation of the results of the experiment.[13]

Some authors, however, reject Duhem's thesis on the basis of a weaker argument, claiming that conclusive falsification is possible when the accompanying knowledge is identical for the two competing theories. Thus, for example, Popper writes:

> Against the view here developed one might be tempted to object (following Duhem) that in every test it is not only the theory under investigation which is involved, but also the whole system of our theories and assumptions—in fact, more or less the whole of our knowledge—so that we can never be certain which of all these assumptions is refuted.

But this criticism overlooks the fact that if we take each of the two theories (between which the crucial experiment is to decide) *together* with all this background knowledge, as indeed we must, then we decide between two systems which differ *only* over the two theories which are at stake. It further overlooks the fact that we do not assert the refutation of the theory as such, but of the theory *together* with that background knowledge; parts of which, if other crucial experiments can be designed, may indeed one day be rejected as responsible for the failure.[14]

Let us note that Popper does not claim here, as he did in *The Logic of Scientific Discovery*, that "Duhem denies [...] the possibility of crucial experiments, because he thinks of them as verifications, while I assert the possibility of crucial *falsifying* experiments."[15] Despite his earlier position, he now admits that the logical asymmetry of the schemes of falsification and verification does not by itself insure the conclusive character of the actual falsifying procedure.

Let us note further that the second argument used by Popper in the cited fragment of *Conjectures and Refutations* indicates clearly that the author does not assume that the background knowledge is certain; he does state after all that it might be rejected in the future. But if so, then the first argument cited in this fragment is not sufficient to reject Duhem's thesis. One can see then that the identity of the background knowledge (if it contains false assumptions) in the two cases does not exclude the possibility of rejecting a true theory and accepting a false one: the conjunction of a true theory and false premises might lead to a false empirical prediction (despite the fact that the theory is true), while the conjunction of a false theory and false assumptions might result in a true prediction. In short: a crucial experiment is not conclusive even when the background knowledge is identical for the two competing theories. In order for it to be considered conclusive, it would have to be assumed that the background knowledge is true. Otherwise, a conclusive elimination of a false theory is not possible.

The entire argument of Popper contained in the cited fragment betrays in my opinion the shakiness of his views on this issue. First, to say that what is being tested is not a theory, but a theory together with its accompanying background knowledge, in no way contradicts Duhem's thesis of the impossibility of the decisive falsification of an isolated theory. It is in fact rather an acceptance than a rejection of Duhem's position, and Duhem does not overlook anything here. Secondly, it would be difficult to call an experiment crucial if, when on its basis we reject a theory together with its background knowledge without knowing which was falsified, we have no grounds for admitting a competing theory accompanied by the same background knowledge. Such an experiment does not give us any indirect corroboration of a competing theory. Thirdly, if we accept the possibility that the background knowledge will be modified in the future, then falsification is not conclusive for this very reason. And fourthly, if we were to demand that the accepted background knowledge be subjected systematically to crucial experiments, we would inevitably end up in an infinite regress.

It does not seem necessary to prove that accepting all of the background knowledge as unquestionable (*de facto*, not by convention) cannot be justified. And I know of no author who would accept such an idea.

If we cannot claim the existence of a methodological (rather than logical) asymmetry between falsification and verification, if we are unable to defend the idea that falsification can be decisive when background knowledge is identical for the competing theories, and if we are unable to accept background knowledge as a whole, then we are left with only one possible line of defense for the thesis that crucial experiments are possible. This possibility demands an acceptance of the idea that it is not all of our knowledge, but only some fragment thereof which is involved in a crucial experiment; and that if these assumptions are sufficiently confirmed by independent crucial experiments, then conclusive falsification is again possible.[16]

It seems to me that such argumentation is based on two dubious epistemological premises. It assumes not only that (a) we are able to isolate a certain fragment of our knowledge from all the rest, but also that (b) we can identify all the assumptions which support the fragment we have isolated. In other words, this argumentation assumes not only that we are able to ascertain what assumptions are not included in the tested fragment, those which have no influence on the results of the experiment we are interested in, but moreover that we can list all the assumptions on which the results do depend and which do belong to the isolated fragment. The question as to whether these assumptions are generally confirmed, certain, or dubious cannot even be considered if we do not first accept premises (a) and (b). Let us examine these premises more closely.

Ascertaining which assumptions are irrelevant for the results of a crucial experiment is equivalent to the exclusion of certain explicatory hypotheses, or to the decision that they have been conclusively falsified. This part of the argument is then flawed by a *petitio principii* error in so far as it assumes that crucial experiments are possible, asserting that some hypotheses have already been decisively falsified (falsified, not just rejected). By analogy, it is an argument of the same sort as the argument that a given system is perfectly isolated if we are able to list all the factors which have no influence upon it. In his *Dialogue Concerning the Two Chief World Systems*, Galileo, considering the problem of tides, has Sagredo claim that it is thoughtless nonsense to suppose that tides can depend on the motions of the moon. It would be interesting to know what motives induced Galileo to exclude *a priori* this hypothesis as a possible explanation of tides. (We can speculate, for example, that considerations such as opposition to astrological notions which would explain events on earth by reference to heavenly phenomena played a role.) Still, it does not seem to me that anyone can plausibly claim that by isolating some fragment of knowledge and setting it apart from those assumptions on which its truth could not depend, he can be certain of avoiding Galileo's mistake. So much for the first premise.

The second premise, according to which we can list all the assumptions which are in fact connected with a given fragment of knowledge which we are testing, implies that our knowledge is transparent to us. It does not take into account the seemingly obvious fact that some of our convictions appear to us not as assumptions but as obvious facts, and that as long as they remain unquestioned we can remain unaware of their presence. It is the case that classical mechanics was based, among other things, on assumptions such as the non-existence of the curvature of space, simultaneity, the congruence of measuring apparatus despite movement in space, the isochronism of clocks in transport, etc. These assumptions were either treated as obvious or remained completely unarticulated and were accepted without reflection. A substantial intellectual effort was required before these assumptions were discovered, analyzed and criticized. In any case, before the formulation of the theory of relativity, scientists as a rule were unaware that they were accepting such assumptions, since before Einstein nobody had any idea that it could be otherwise, or that if it were otherwise, then the validity of their accepted theories would become problematic.

It seems that if we were aware of all the assumptions which we accept and wanted merely to check them all, not to mention verifying or falsifying them conclusively, then scientific activity would become altogether impossible. It is not proper thus to make a virtue of necessity. It is certainly true that in practice we can control only some fragments of our theoretical knowledge, and not the entirety of this knowledge, and that in order to do so we must assume that these fragments are "well and appropriately isolated," so that by isolating some of them we do not ignore anything which could decide the positive or negative outcome of our testing procedures. We have to accept this assumption if we want to investigate anything. Yet we should not believe that this assumption renders us immune to error. It is also certainly the case that if before performing a test we always waited to identify all the assumptions and convictions that are actually involved in the fragment of knowledge we would like to test, we could never begin our work. But it is not legitimate to conclude from this that tests can be conclusive.

The arguments which I have cited here, which I consider essential in the entire discussion about the possibility of crucial experiments, are directed against those who claim (a) that no hypothesis can be tested alone, but who then stop halfway and add (b) that some limited fragment of knowledge or a set of hypotheses can be conclusively tested in isolation.[17] Such a position seems to me to be inconsistent, since all the arguments against thesis (a) are equally valid against thesis (b). It is worth remembering that no cognitive procedure is possible without this assumption, but precisely because of this, no such procedure can be conclusive, even if we accept the indisputable nature of empirical evidence. I can understand that one could adopt an atomistic position on this issue and consider the arguments against thesis (a) as unconvincing. I also understand that one could accept this thesis and accept holism. It seems to me, however, that *tertium non datur*: in other words, a compromise solution to this theoretical problem is impossible.

5.

If one introduces no methodological postulates either to forbid completely all modifications of background knowledge (that is, forbid all *ad hoc* hypotheses), or to indicate which modifications are permissible and which are not, then Duhem's thesis is a sufficient basis for acceptance of the statement that in any falsifying experiment, the scientist can just as well defend the theory which is being questioned as reject it and accept its competitor; in other words, he possesses no unequivocal methodological selection criteria.

Although Duhem's thesis does not guarantee that the scientists will succeed by choosing any of the alternatives, it is obvious that no methodology can answer the question of how many attempts to save a theory that has been questioned by experiment will be necessary before the continuation of such attempts to save the theory will violate the rational methods of scientific practice. From the point of view of the controversy over the existence of unequivocal selection criteria, the acceptance of Duhem's thesis is decisive, even if its more radical variant—that of Quine—is unacceptable, that is, even if it were not always possible to specify modifications of background knowledge to eliminate the inconsistencies between a theory and an experiment.

The fact that such modifications are sometimes possible, which nobody denies, is sufficient to warrant considering the search for such a possibility, no matter how long it might take, as equally justified methodologically as the abandonment of the theory in favor of its competitor. Still, despite the fact that the truth of Duhem's thesis is a condition sufficient to defend a position according to which a methodology cannot provide unequivocal selection criteria, Quine's thesis is worth examining a little more closely:

> Even a statement very close to the periphery [that is, subject to direct sensory control,—S.A.] can be held true in the face of recalcitrant experience by pleading hallucination or by amending certain statements of the kind called logical laws. Conversely, by the same token, no statement is immune to revision. Revision even of the logical law of the excluded middle has been proposed as a means of simplifying quantum mechanics; and what difference is there in principle between such a shift and the shift whereby Kepler superseded Ptolemy, or Einstein Newton, or Darwin Aristotle?[18]

> A recalcitrant experience can, I have urged, be accommodated by any of various alternative reevaluations in various alternative quarters of the total system; but [...] our natural tendency to disturb the total system as little as possible would lead us to focus our revisions upon these specific statements [which have been directly challenged by experience—S.A.]. These statements are felt, therefore, to have a sharper empirical reference than highly theoretical statements of physics or logic or ontology. The latter statements may be thought of as relatively centrally located within the total network, meaning merely that little preferential connection with any particular sense data obtrudes itself.[19]

In his polemics against this position, Grünbaum uses two arguments. He says first that it is of course always possible to find such trivial modifications of background knowledge (or of other statements in the system—to use Quine's terminology) which will logically guarantee agreement between the results of experience and theory with assumptions thus modified. This, however, does not mean that in every situation in the empirical sciences there exists such a non-trivial set of assumptions A' which will guarantee that the conjunction $H \cdot A'$ will imply the occurrence of $\sim E$, that is, of the result which undermines $H \cdot A$.[20] Secondly, Grünbaum argues that the truth of the non-trivial thesis of Quine has not been shown.[21]

I believe Grünbaum is absolutely right in distinguishing the trivial from the non-trivial version of the thesis under discussion. The thesis is methodologically interesting only if we impose some formal restrictions on the admissibility of the modifications of background knowledge: at the very least, excluding those whereby the modified assumptions alone, without the help of the hypothesis under test, would imply the occurrence of the results obtained from the experiment.

It seems to me, however, that Grünbaum asks for too much when he demands that the supporters of Quine's thesis demonstrate that there always exists a non-trivial set of assumptions A'.

First of all, the proof of the existence of non-trivial A' requires the formulation of conditions sufficient for deciding that A' is at least formally non-trivial. At the same time, Grünbaum himself admits that he cannot "give a formal and completely general sufficient condition for the non-triviality of A'."[22] In a later work he even admits that such conditions cannot be given, since it is impossible to formalize the concept of an *ad hoc* hypothesis in a satisfactory manner.[23]

Grünbaum is doubtless correct when he argues that the non-trivial version of Quine's thesis does not follow logically from Duhem's thesis; it will always be possible to find such non-trivial assumptions to save the theory. But in demanding a proof of this thesis, he is asking for the impossible, since it is impossible to formulate conditions of non-triviality.

Moreover, Grünbaum's demand appears too strong for yet another reason. Quine's thesis, even if not proven, is methodologically interesting as soon as we cannot exclude the possibility of finding a non-trivial modification of background knowledge. If, in the expression "there exists a non-trivial set of assumptions A' which allows us to maintain the hypothesis H in face of any experimental result," the word "exists" is supposed to mean that such a set of assumptions is known in every conceivable situation in the empirical sciences, then the thesis under discussion would evidently be false. If I understand Quine's thesis correctly, what it states is not that such a set is known, but that given sufficient ingenuity among scientists it could be found.

When we realize how many different assumptions—theories from other disciplines, ontological and epistemological beliefs, logical laws—are involved in every attempt to falsify a scientific theory, then the supposition that such a set of assumptions can be found appears probable. Grünbaum is right that nobody has proven this to be the case, but—we can respond —no one has disproven it either. Moreover, it appears that nobody can either prove it or disprove it for reasons which Grünbaum himself indicates and accepts: it is impossible to formalize conditions for the non-triviality of such a set of assumptions. Under these conditions, it is pointless to demand proof from either side of the debate, and the entire controversy becomes an expression of different philosophical views about science, or more exactly, about the relationship between science and reality.

Quine's views are doubtless linked with an instrumental attitude towards the theory of science. He writes:

> As an empiricist I continue to think of the conceptual scheme of science as a tool, ultimately, for predicting future experience in the light of past experience. Physical objects are conceptually imported into the situation as convenient intermediaries—not by definition in terms of experience, but simply as irreducible posits [...] The myth of physical objects is epistemologically superior to most in that it has proved more efficacious than other myths as a device for working a manageable structure into the flux of experience.[24]

There is, however, no necessary connection between accepting Quine's thesis and adopting an instrumentalist position in the theory of science. Not every instrumentalist must accept the thesis, and not everyone who accepts the thesis must share Quine's instrumentalist views. The thesis implies only that there are many conceptual systems in which our knowledge of reality might be articulated. It does not follow that each of these systems might have only an instrumental value, and cannot be judged at all in terms of its substantive relation to reality. I would say that this is rather the relativist thesis. It does admit that there might be more than one conceptual system privileged to give an account of reality, and claims that the selection criteria between such systems cannot be codified methodologically.

We must thus conclude that the dispute about the truth of Quine's thesis is undecidable. Although personally I am inclined to believe that the thesis is correct, I will not appeal to it later, since in order to argue for the position I am trying to present here, it is enough to accept Duhem's weaker thesis that it is impossible to falsify conclusively any delimited fragment of our knowledge.

6.

I argued above that the acceptance of Duhem's thesis is sufficient to justify the view that no methodology can formulate unequivocal selection criteria *unless* it introduces a rule banning all modifications of background knowledge or indicating which modifications are permissible and which are not. This raises the question of whether it is possible to formulate such rules in a satisfactory manner.

It is easy to show that a methodological rule forbidding *all* modifications in background knowledge in order to save a tested theory from falsification is unacceptable. Such a rule is neither respected nor could it be respected in scientific activity.

The fact that stones fall vertically, that houses and fortresses do not disintegrate into ruins, that wine does not splash from a pitcher, and that no wind blows constantly from the east, could serve just as well as evidence against the Copernican theory as against Aristotle's physics. It was possible to reason that given these facts the heliocentric hypothesis in unacceptable; but it was also possible to ask how all these facts are possible if the earth turns.

Similarly, the fact that the orbit of Uranus does not agree with Newton's theory could be interpreted as a sign of the imprecision of the theory, but it could also serve as grounds for the hypothesis that the assumption about the number of planets in the solar system is incorrect, and that in fact there exists an as yet undiscovered planet which disturbs the orbit of Uranus (such a hypothesis was proposed in 1821 by Brouvard, and in 1866 Leverrier discovered Neptune).

In both of these cases the removal of the anomaly involved a revision of background knowledge: in the first case, a revision of the very general physical theory of Aristotle, which made it impossible to reconcile the observed facts with the heliocentric hypothesis; while in the second case, the modification was introduced in a specific claim describing the so-called boundary conditions of the system to which Newton's theory was applied. And yet, in the case of the anomalies in the orbit of Mercury, the hypothesis of the existence of another unknown planet (Pluto) which was presumed responsible for these disturbances was not confirmed, and the anomaly was removed only after the introduction of the theory of relativity, that is, by a revision of Newton's theory.

It is thus impossible to accept a rule which would forbid all modification of background knowledge, since in effect this would mean the defense of older, and not necessarily better, theories. Given such a rule, in every case of competition between two theories, the winner would be the theory most in agreement with the knowledge already sanctioned by tradition and functioning precisely as background knowledge. Were this rule to be observed, neither local nor global revolutions in science would be possible.

It is more complicated to evaluate a rule which would exclude only certain kinds of, rather than all, modifications of background knowledge. Banning *ad hoc* hypotheses, it would designate by this term only certain types of modifications of previous knowledge.

Following Lakatos in distinguishing among dogmatic, naive, and sophisticated falsificationism, we could say that:

a) Dogmatic falsificationism rejects Duhem's thesis at the limit and maintains that the falsification of a theory can be conclusive and can prove the theory's falsity.

b) Naive falsificationism accepts Duhem's thesis but assumes that falsification can be definitive on the basis of a rule forbidding modifications of background knowledge. In this case falsification does not prove that the theory is false, but would demand its elimination at least temporarily.

c) Sophisticated falsificationism introduces a further liberalization of the criteria of falsification, admitting some modifications of background knowledge; but it maintains the view that by distinguishing permissible from impermissible modifications it is possible to formulate unequivocal methodological selection criteria.

If I were to classify the positions of specific authors in terms of these categories, I would agree with Lakatos that Popper's positions evolved from dogmatic falsificationism in the 1920s (traces of which can still be found in *The Logic of Scientific Discovery*) through the naive falsificationism which dominated the *Logic*, to several versions of sophisticated falsificationism which have been articulated since the 1950s (in *Conjectures and Refutations* and later works).[25] These views were later developed—often without the master's approval, though with frequent appeals to his work—by Popper's students: Lakatos, Watkins, Zachar, Musgrave, Worrall and others. In any case Popper never rejected the view that "*criteria of refutation* have to be laid down beforehand: it must be agreed which observable situations, if actually observed, mean that the theory is refuted."[26]

The idea of the methodology of research programmes, presented mainly by Lakatos and defended after his death by many authors from Popper's school, constitutes a real development of "sophisticated falsificationism." Evaluating this discussion from the perspective of over a dozen years, I think I was correct in my earlier characterization of the methodology of scientific research programmes as an attempted counter-reformation in view of the positions of Kuhn, Feyerabend, Toulmin, and others.[27]

We need here to pose two questions: first, can one really formulate methodological rules allowing us to specify which modifications introduced in order to save a theory from experimental falsification are acceptable and which are not? and secondly, should these rules apply to individual theories or to research programmes? The difference between naive and sophisticated falsificationism concerns primarily this second question, while on the first both positions basically agree. Let us address these questions in turn.

All the authors who have attempted to define which modifications introduced in order to save a theory from experimental falsification are acceptable (Popper, Grünbaum, Laudan, Leplin and Lakatos) have at least agreed that it is impermissible to introduce modifications which have no (independent) empirical consequences apart from the one which constituted an anomaly with respect to the tested theory. In other words, a modification of background assumptions, if it is not to be an *ad hoc* hypothesis in the pejorative sense of this term, must lead to a situation in which the old theory with its new assumptions can explain something more than (a) the facts explained by the old theory and (b) the observed anomaly. Such an understanding of inadmissible modifications corresponds to what we usually have in mind when speaking of *ad hoc* hypotheses. Illustrating this with an example, we could say that if the hypothesis of the existence of the neutrino, introduced by Wolfgang Pauli in order to explain a violation in the law of conservation of energy in beta decay, did not explain anything beyond this fact, it would have to be dismissed as an *ad hoc* hypothesis introduced only to preserve the existing physical theory (the law of conservation of energy) in an inadmissible manner.

Thus, the basic problem of distinguishing between admissible and inadmissible modifications of background knowledge would be reduced to the issue of how to distinguish a hypothesis which has independent empirical consequences from one which does not. Either we are able to provide some formal characteristics of hypotheses which are admissible in this sense, or the decision about whether or not a given hypothesis is admissible depends not on whether it has independent empirical consequences, but on whether *such consequences are known to us*. In the first case, the distinction would be based on the methodological characterization of an auxiliary hypothesis; in the second it would be historically conditional, relativized to the current state of knowledge on the basis of which the decision would have to be made about whether or not the modification is acceptable. In the first case, methodology would be able to formulate selection criteria, since it asserts that in the case of contradictions between the theory under test and the result of the experiment, the theory in question can be saved only by hypotheses having independent empirical consequences; in the second case, it would be impossible to cite any criterion, since the fact that at the moment we know of no such independent empirical consequences does not mean that such consequences do not exist. In this second case, the question of "what is an *ad hoc* hypothesis?" remains without a formally correct answer. We could say only that an *ad hoc* hypothesis *is a hypothesis whose independent empirical consequences are at this point in time unknown to us.* What today is considered an *ad hoc* hypothesis might appear differently tomorrow. A prohibition against the introduction of such hypotheses could not be methodologically sanctioned.

Let us note incidentally that the concept of the *ad hoc* hypothesis accepted and discussed here is logically the weakest, since one can demand that an acceptable hypothesis have not only (independent) empirical consequences, but moreover, that these consequences not contradict other empirically confirmed theories or—even more—that they be experimentally confirmed.

In a paper devoted specifically to this problem, Grünbaum analyzes various possible definitions of *ad hoc* hypotheses in order to conclude that it is impossible to give any formal definition of the term "independent consequence."[28] This means that we cannot formalize the concept of the *ad hoc* hypothesis and the conditions for admitting such hypotheses in order to save scientific theories threatened by experimental results.

A similar conclusion was also reached by Hempel, who stated, "There is, in fact, no precise criterion for *ad hoc* hypothesis."[29] Hempel finally accepts Duhem's view that theories fall when as a result of introduced modifications they become excessively complicated.

It is perhaps worth noting that both of the cited authors were previously very far from the view that a definite acceptance (Hempel) or definite falsification (Grünbaum) is impossible.

The history of science seems to confirm this negative conclusion. At the time when Leverrier's hypothesis of the unknown planet disturbing the orbit of Uranus was introduced, it was without doubt an *ad hoc* hypothesis, even if this term is understood in its weakest form.[30] This was also the character of Leverrier's hypothesis about the existence of the planet Pluto, which was adduced to explain the anomalies in the orbit of Mercury. Both hypotheses had an identical methodological status—neither had any independent empirical consequences known at the time, not even uncon-firmed ones. The first turned out to be correct, while the second did not find confirmation. Both were designed to save Newton's theory from falsification by the results of astronomical observation, and there are no methodological rules to indicate why Newton's theory should have been saved by the search for an as yet unknown planet in the first case, while in the second case it should have been revised (the anomalies of perihelion of Mercury are explained by relativity theory). There are more such cases in the history of science.

Thus, the weakest conclusion which we can draw from this analysis is that *crucial experiments are impossible in science; Duhem's thesis is correct, and there are no methodologically privileged methods of saving theories from falsification. It is possible* ex post *to decide what* de facto *became a crucial experiment, but it is not possible to determine* de jure *what could not have been considered as such an experiment, even if the choice that was made turned out to have been correct.*

7.

Lakatos's methodology of research programmes (as against Popper) accepts this conclusion regarding scientific theories but attempts to avoid the consequence that no methodology is able to formulate unequivocal criteria of choice. The basic idea is to treat the development of knowledge not as a succession of theories, but as a realization of successive research programmes. The name chosen appeals without doubt to the Popperian idea of metaphysical research programmes which was developed by Agassi.[31] By including metaphysical ideas in the core of the programme, the methodology of research programmes undermines in effect the entire Popperian idea of the demarcation of science from metaphysics. This did not, incidentally, prevent Lakatos from claiming that his idea was a continuation and improvement of Popper's philosophy.[32]

According to Lakatos, methodological rules have to indicate what relations should obtain between successive theories within the bounds of a programme which is being realized, so that this programme remains progressive, that is, so that it develops in a manner such that each suc-cessive theory explains everything that its predecessor did plus at least some of the anomalies with which the old theory could not cope. The essential change from Popper's falsificationism comes with the idea that as long as there is no theory which meets the condition of "progressiveness," the previous theory is not to be considered falsified even if many empirical facts which do not agree with it are known. It is said that in such a situation it is rational both to search for a new theory and to attempt to save the old one, since the negative heuristics of a research programme prohibits the use of the *modus tollens* with respect to the assumptions of a research programme which belong to its core. However, the moment such a new theory meeting the condition of progressiveness comes to be formulated, it becomes irrational to defend the old theory.[33]

In other words, in cases of conflict between experience and theory, Lakatos offers theory time to prove its fitness to survive, and treats as rational all attempts to save the theory in conformity with the basic assumptions of the programme. Yet as soon as a new theory appears on the scene which fulfills the condition of being progressive, the decision is made: the programme turned out to have been progressive (it gave birth to a new, better theory); while the old theory has by the same token been falsified and should be rejected. The accepted selection criterion speaks unequivocally (within the framework of a given programme) for the new theory.

The solution to the problem of selection criteria within the framework of an accepted programme is based, explicitly, on the decision that there is a "court of appeal": the research programme with its positive and negative heuristics; and this court decides the repertoire of acceptable changes (prohibits revising inviolable assumptions, indicates how to construct auxiliary hypotheses which would agree with metaphysical assumptions, etc.). The methodological criterion of progressiveness is the tool for choosing from this repertoire.

However we evaluate this modification of the classical falsificationist position, (a modification which rejects both the criterion of the demarcation of science from metaphysics and the requirement that theories should be rejected as soon as experience speaks against them), there can be no doubt that the problem of the criteria of selection re-emerges in exactly the same form when it is necessary to choose between alternative programmes.[34] It is easy to imagine a situation in which all attempts fail to remove an empirical anomaly within the framework of a given programme, so that a new theory progressive with respect to the old one does not make its appearance. Should scientists then continue their search for such a theory, or should they rather modify the assumptions which until then, within the framework of the programme, were considered inviolable, that is should they search for a new programme? The first course would be consistent with Lakatos's view that the old theory has not been falsified until a new one appears; the second course of action would also be rational, however, because it is impossible to exclude the possibility that a change in the research programme might lead to progress. A methodology of research programmes does not in any way solve this problem.

In a private conversation Lakatos told me that like many other critics I have not understood his position. He claimed that his methodology does not aim at all at the solution of the problem of transition from one research programme to another (either normatively or descriptively), but that it is meant to supply rational rules for evaluating theories and programmes *ex post*.[35] In this situation, however, it is difficult to understand where he disagreed with Kuhn and Feyerabend. It also becomes understandable why Feyerabend dedicated his book *Against Method* "To Imre Lakatos: Friend, and fellow-anarchist," since he proves that in fact Lakatos accepts the principle of "anything goes," that is, he abandons all methodological selection criteria. The danger that Lakatos's position can be so interpreted also worries Musgrave, who attempts to save the methodology of research programmes from this consequence, which was undoubtedly contrary to Lakatos's intentions.[36]

The position of Lakatos in this matter was clearly ambivalent, though it is difficult to doubt that it was conceived in opposition to the views of Kuhn and Feyerabend. I believe, however, that given Lakatos's view of science, this ambivalence was inescapable. He claims to be providing "the rules for the 'elimination' of entire research programmes," but—as Feyerabend correctly notes—he places the word "eliminate" in quotation marks, which makes the sentence unclear, while in another place he claims that "one may rationally stick to a degenerating programme until it is overtaken by its rival *and even after*."[37]

It is impossible to eliminate this ambiguity, since while a research programme constitutes a court of appeal for a theory developing within the framework of such a programme, no such court exists for the programmes themselves. There is no way to establish how many unsuccessful attempts to save a degenerating programme are necessary before it becomes advisable to abandon it and to search for a new one; and even when a competing programme already exists, one cannot exclude the possibility that the n-th attempt to save the old programme will end in success, while the competing programme in fact faces degeneration. In other words, the methodology of research programmes faces exactly the same difficulty as that encountered by Kuhn in his attempt to explain changes of paradigms. There is, however, at least one important difference: Kuhn believes that this difficulty cannot be solved by methodological means, and that to explain transitions from one paradigm to another it is necessary to go beyond methodology; while the entire system of Lakatos was supposed to

demonstrate that this conclusion is unjustified and to show the possibility
of explaining and reconstructing such changes in purely methodological
categories. In this sense, the methodology of research programmes certainly has not solved the problem it was intended to solve.

If the terminology employed by the participants in this controversy
were used to characterize the controversy itself, one could argue that we
are facing here two distinct paradigms or research programmes proposing
how to conduct philosophical reflection on science. The situation is such
that while Kuhn's programme has shown that the mechanism of the development of scientific knowledge cannot be reconstructed purely in terms of
methodological rules, and that on the basis of the criteria of rationality
identified with these rules it is necessary to conclude that the process of
scientific development is not purely rational, he is nevertheless unable to
explain this process of change in categories which he considers appropriate,
or at least he is unable to do so in a satisfactory manner. In this situation
it is of course possible to undertake new attempts to save the old purely
methodological programme, and I have described these attempts "the
counter-reformation." It is equally possible to undertake efforts intended
to improve the new programme. It is not difficult to note that the reflections offered here are directed towards this second possibility.

8.

Before ending our reflections on the question of selection criteria, let
us consider for a moment what we are to understand by Lakatos's claim
that the methodology of research programmes does not provide any
instructions about how to act, and constitutes only a tool for an *ex post*
evaluation of theories and research programmes. How are we to understand this claim, if we assume that it is not to be understood as an expression of Lakatos's agreement with the very position against which the
methodology of research programmes was designed to serve as an antidote,
which stated that a methodology cannot provide conclusive selection
criteria. What then are we to understand by Lakatos's claim that "philosophy of science supplies a normative methodology," if it not only does
not formulate any directions, but also refrains from offering such directions
to scientists. Lakatos himself explains this as follows:

"This is an all-important shift in the problem of normative philosophy
of science. The term 'normative' no longer means rules for arriving at
solutions, but merely directions for the appraisal of solutions already there.

The methodology is separated from *heuristics*, rather as value judgments are from 'ought' statements."[38]

It is of course true that value judgments and judgments concerning what ought to be done are not the same. It is also true that value judgments are not as a rule spoken disinterestedly, but uttered so as to induce others to accept certain definite norms of behavior. When we say that murder is wrong, we do so, among other things, in order to convince others that they should accept the norm of "thou shalt not kill." The normative character of ethics is not reduced to the claim that "murder is wrong, but you can do as you wish, since from the claim that 'murder is wrong' it does not follow that you are not supposed to do something wrong." The normative character of the Lakatosian philosophy of science is limited precisely to this. The view he expressed can be paraphrased: "The philosophy of science provides rules for the rational evaluation of theories or research programmes, but when you are engaged in scientific research, you need not follow these rules and you need not behave rationally." How can one reconcile this understanding of the normative character of methodology with the simultaneously expressed opinion that this is an ethic of cognition and "it offers models of scientific honesty."[39] And why should philosophy of science be practiced then at all? For whom or for what is it useful? Is its only task really just to supply historians with models of evaluation for what they ascertain when they study the history of scientific cognition?

One cannot disagree with Quinn, who argues that "it is hard to see what point methodological appraisals of scientific theories could possibly have is such appraisals were completely decoupled from heuristic advice, if they had no practical force."[40]

In one of his earlier works, a polemics with inductivist methodology, Lakatos himself wrote in very much the same spirit: "the appraisal of any finished product is bound to have decisive pragmatic consequences for the method of its production. Moral standards, by which one judges people, have great pragmatic implications for education, that is, for the method of their production. Similarly, scientific standards by which one judges theories, have grave pragmatic implications for scientific method, the method of their production."[41]

So, how do we distinguish methodology from heuristics?

We find ourselves in a troublesome situation in which methodology would have to limit its competence to the evaluation of the past according to the accepted supra-historical criteria of rationality—in which it would specify why it was rational to accept Einstein's theory, but would be unable to formulate any directions to scientists requiring, for example, that they continue Einstein's program rather than Newton's.[42] One means of escape from this situation is to acknowledge that although *methodology gives no directions to individual scientists, it does provide directions to the scientific community as a whole.*

So, for example, Grünbaum writes:

> It is important to decide that heuristically it might be rational for some scientists to continue research on a quasi-falsified hypothesis. Methodologically, however, it would be irrational if all scientists followed such a hypothesis. [...] A heurstics which I propose to link with my notion of quasi-falsifiability allows us, I believe, to avoid both of these extremes: the Scylla of deprecating methodological rationality in the name of historical accuracy, and the Charybdis of sacrificing both historical accuracy and the rationale of all scientific heterodoxy at the altar of methodological correctness. [...] Since scientific inquiry is conducted by a *community* of scientists, [...] practices that would indeed be irrational if adopted by that community as a whole or by a majority of it need not necessarily be irrational when only a certain gifted minority engages in them.[43]

Although Musgrave does not share Grünbaum's opinion about the nature of this quasi-falsification, but believes instead (as did Lakatos) than even such a quasi-falsification is impossible, his opinion on the problem at issue here is identical.[44] Musgrave writes:

> The abdication of the Lakatosian methodologist is complete. He promised to hand down judgement on the rationality or otherwise of pursuing some research programme; now he is ending up by trying to supply rules to appraise research programmes on the rationality of pursuing which he will not pronounce. The Lakatosian methodologist cannot claim any longer to be a guide of scientific life in any relevant sense. [...]
>
> If the methodology of research programmes (or indeed any other methodology) is to provide advice or directives, then these must be addressed to science, or if you prefer, to the community of scientists, *as a whole*. Such community-directed advice would not forbid an individual (or even a group of them) from persisting with a degenerating programme when a progressive rival is available, or from working within a budding programme which at the time is inferior to an established rival. It *would* forbid wholesale persistence with a degenerating programme, or premature mass-conversion to a budding one (both of which would have to be explained on "externalist" grounds). We cannot condemn Priestley for his die-hard adherence to phlogistonism; but we could condemn the community of late eighteenth century chemists had they all done the same.[45]

But what does it mean to say that heuristic directions based on methodological evaluations are addressed to the "whole community" of scientists, rather than to individuals? I can find no answer to this question other than that they are directed to the institutions organizing and financing scientific investigations. Who else could otherwise be using them? Who else has the power to decide how numerous the "gifted minority" is, and how gifted it has to be? But institutions are also unable to implement these directions other than with the help of its policy to support specific research programmes, that is, above all, by the distribution of the material resources which they have under their control. In this conception, a *methodology would not be the basis of cognitive activity but of science policy*. It would not specify rational selection criteria, but rather the rational criteria for the investment of resources in research programmes in risky situations.

The abdication of Lakatos's methodologist is, however, even more complete than Musgrave realizes, and so the correction he has introduced is not worth much. In this situation, the methodologist gives up his ambition to formulate the rules of the ethics of cognition. He can say to the scientist nothing more than that he should not present a degenerating programme which he does not want to abandon as a programme promising success. He aspires instead to become an advisor to science policy managers, regardless of whether they are working in the private or the governmental sphere. (Incidentally, in this respect again it is difficult to treat Lakatos as a continuator of Popper's philosophy of science.) Although the methodologist cannot say that a stubborn adherence to a degenerating programme is irrational, by directing his remarks to "the scientific community as a whole" (that is, to institutions), through them he can limit the resources which scientists receive for this research. In a limiting case he can contribute to the total denial of funds for particular research.

Lakatos himself seems to have been close to this idea of Musgrave when he wrote: "Editors of scientific journals should refuse to publish their papers [i.e., the papers of the supporters of a degenerating programme— S.A.] [...] Research foundations, too, should refuse money."[46]

Before science was professionalized, when a scientist was still basically independent of any institution in the choice of his research programme, such directions addressed to "the scientific community" did not in reality have an addressee able to implement them, and would signify only an invitation to tolerate "weirdos." At present, when without technical equipment and instrumentation, without colleagues and collaborators, and without financial support, an individual scientist has no chance to compete alone with any institutionally supported research programme, such directions put the individual from the start in a losing situation, even when they are implemented in a "liberal" manner, that is, even if we assume that institutions will be intelligent enough not to risk everything they have on a single bet and will simultaneously support different programmes (this is surely what Lakatos, Grünbaum and Musgrave would advise them to do). But even a liberal policy on this issue, that is, institutional support for programmes according to the methodological evaluation of their chances of success, condemns a degenerating programme. From the fact that its chances of success are smaller, it does not follow that it can achieve such success by receiving less institutional support, or by receiving no support at all. Institutional policy—if the institutions involved are not charitable organizations—must, in this respect, be conservative.

Long ago, when scientific activity ceased to be just an intellectual adventure and became a profession and a business proposition, science policy, even a policy conducted according to the direction of the methodology of scientific research programmes, would lead—at least if generally adopted—to the abandonment of some unsupported or weakly supported programmes for non-substantive reasons despite their small (but finally undecidable) chances of success. What is not and cannot be accomplished by methodological arguments appealing to quasi-falsification or rationality, what cannot be disqualified as irrational behavior by any methodological rules, will take place "by itself" as a result of institutional pressures and sociological mechanisms working in the professionalized scientific community.

It seems quite symptomatic that our discussion about the existence of conclusive criteria for the selection of theories which would constitute a rationale of the development of science leads us finally to search for guarantees of this rationality in the behavior of institutions. Institutions are to guarantee what methodology cannot. To claim today that methodology offers no heuristic rules to the individual scientist, but rather addresses itself to the scientific community as a whole, means basically that we agree that decisions concerning the choice of research programmes will in practice be subordinated to quite different criteria of rationality that those which are supported by the methodology. They will be subordinated to the criteria of rationality of the institutions making these decisions.

In order to be rid of all illusions in this matter it is enough to reflect for a moment on factors other than the methodological fruitfulness of the programme, that is, factors such as the expected extra-cognitive returns, financial considerations, competition etc., which influence and must influence institutional decisions. In these types of decisions, as Lakatos himself agrees, one has to appeal to common sense.[47] It is difficult to doubt that this is precisely what is happening in contemporary science. But if this is really the case, then addressing heuristic directions to the scientific community as a whole amounts to an assent to the subordination of decisions concerning the choice of research programmes to the rational criteria of "common sense" as interpreted by these institutions. Musgrave seems not to realize that the abdication of the methodologist means here a consent to replace his criteria with quite different criteria. The rationality of the scientific method ceases to be the guarantor of scientific rationality and comes to be replaced by something quite different, namely the rationality of the functioning of these institutions which make decisions in the name of the entire community or, possibly, some part of that community. It can be said that the sociological problems of the development of science, programmatically expelled from philosophy, return by the back door. A philosophy of science which limits itself to methodological issues can neither solve these problems nor maintain its critical stance towards the actual science policy conducted by these institutions. It is to these institutions that it transfers the responsibility for the development of science, legitimizing their rationality or—possibly—announcing its own lack of interest in such issues.

The remaining question is how such a methodology could claim to be the basis of the rational reconstruction of the development of science? What rationality would be revealed by this reconstruction? How can one distinguish between "internal history" and "irrational external history" if, given the lack of any conclusive methodological selection criteria, the selection of theories is made with the help of models of rationality based on "common sense"? It is difficult to expect that methodology would be able to tell us here more than that if this development is rational, then it is not because of supra-historical methodological rules which it attempts to formulate, and which do not supply conclusive rules for the selection of theories, but because of the models of rationality derived from elsewhere which are certainly changeable, just as changeable as are the canons of common sense.

To say that the methodology provides the community of scientists with heuristic directions implies that the abandonment of the thesis that sociological, historical and psychological factors is unimportant in understanding the development of science. The question of how institutions use or could use the rational directions of methodologists remains open. This is certainly a sociological and a historical question. When heuristic directions are no longer directed to an individual knowing subject but to communities, it is no longer possible to avoid sociological problematics. The critique of science becomes inseparable from the critique of scientific institutions and of their functioning. There is reason to expect that rational principles for the functioning of institutions will dominate the rationality of the directions delivered by a methodology, and the development of science will be rational to the extent (and in the sense) that the functioning of the institutions or the social structure to which they belong can be considered rational. What is most important, however, is that from this perspective the problem of rationality assumes new dimensions.

In summary, I can say that if the methodology of research programmes abandons all attempts to formulate the heuristic rules addressed to a scientist, then it does indeed approach the position of epistemological anarchism which was initially its main enemy. It has value only as an evaluative program (but not a reconstructive one) for the history of science on the basis of the models of rationality it accepts. If such directives are aimed at the scientific community as a whole, then instead of supplying rules for the ethics of cognition, the methodology of research programmes supplies only indications for science policy, and—again contrary to its initial assumptions—it cannot avoid sociological problematics if it is to explain the development of science. The firm division between internal and external history of science becomes a fiction.

In this manner we have successively disposed of all the possibilities of defending the thesis that methodology is able to formulate or reconstruct unequivocal, empirical selection criteria for theories (that crucial experiments are possible), criteria which—whether descriptively or normatively —would be able to serve as a model for the mechanisms of scientific development.

Since no fragment of knowledge can be conclusively confronted with experience in isolation from other knowledge; since it is impossible to assume that background knowledge (whether it belongs or does not belong to this fragment) is indubitably correct; since it is impossible to forbid all modifications of background knowledge in order to protect the theory under test; since on the basis of a methodology it is impossible to state generally which modification of background knowledge are permissible and which are not; and since, moreover, heuristic rules directed to the scientific community as a whole are either (under the conditions of non-professionalized science) devoid of an addressee able to implement them, or (in the situation of professionalized science) directed to institutions which act according to criteria of rationality derived from sources other than methodology; then for all these reasons the explanation of the process of scientific development indeed demands that we step beyond the bounds of methodological reflection.

CHAPTER VIII

ORDER AND ANARCHY

1.

The conclusions of the previous chapter might well lead us to wonder how it is possible that despite the lack of conclusive criteria for theory choice, both the number of solved scientific problems and the precision of individual solutions have grown with time. How is this increase possible if, on the one hand, it is claimed that methodological criteria leave a certain degree of freedom, while on the other hand it seems impossible to suspect that given this freedom, scientists will toss coins to make their decisions, and that—in addition—luck has so often been with them.

In considering this issue it is necessary first of all to state that it is one thing to believe that methodology does not and cannot supply conclusive criteria for selecting theories or research programmes, that is, to believe that it is impossible to explain the historical development of scientific knowledge on the basis of such criteria; and it is quite another to believe that the decisions made by scientists in situations requiring choices are completely arbitrary, unconditioned by any objectively presentable factors. In order to avoid misunderstandings in this matter, I will repeat that I am defending only the first of these beliefs, and that it by no means implies the second. The claim that methodological rules do not supply unequivocal selection criteria does not mean that such rules have no influence whatsoever on the choices made, nor that it is impossible to name other factors which influence decisions, nor that the only possible solution is to toss a coin. However, this claim does imply that in order to discover what these factors are, it is necessary to go beyond methodological analysis. This is also necessary because—as I have tried to show in the first two chapters—the accepted methodological rules are not supra-historical.

In other words, I am defending two theses: first, that *it is impossible to represent the development of science as a process which always takes place according to the same methodological rules*, since these rules are not immutable but imposed by the historically conditioned ideals of scientific knowledge; and secondly, that *the methodological rules accepted on the basis of a specific ideal of science do not supply unequivocal criteria of theory choice*, so that even a "short range" theory of science has to consider non-methodological factors.

I may be reproached for having failed to justify fully the second of these theses. The arguments used to support it took into account only the methodological criteria relating to the use of experimental results as a basis for theory choice, while the conclusion presented above states more, namely, that *no* methodological rules can supply unequivocal selection criteria. It is certain that scientists choosing between alternative theories can be guided by some *additional* methodological criteria, such as, for example, *theoretical simplicity, precision, internal consistency, coherence with other accepted theories, generality, fruitfulness, operational or practical utility.* Methodological literature devotes much attention to these issues, and few if any philosophers of science would claim that the results of experiment alone, with no consideration of these additional criteria, could supply an unequivocal verdict.

Although it is true that works on scientific methodology have devoted and continue to devote much attention to these "additional" selection criteria, I do not believe that their inclusion could lead to a revision of the thesis in question.

First, it is very doubtful whether any of the additional criteria can be formulated unequivocally. It is certainly not possible to do so with criteria such as simplicity, generality, or theoretical fruitfulness.

Secondly, and this is most important, it is impossible to rely on all these criteria together, since they can come into conflict with one another. The choice of the simplest theory, even if it were possible to formulate unambiguous criteria of simplicity, need not necessarily mean the choice of the most fruitful or the most general theory. The most general theories are not necessarily the most precise or the most operational. So even if it were possible to formulate these criteria unequivocally, in a manner which would exclude a certain arbitrariness in their application, and even if one could use them to indicate which theory is simpler, more general, more operational, or cognitively more fruitful, there would still remain the freedom to prefer some criteria over others, that is, the problem of their hierarchy. It is certain that no methodology would be able to justify fully any such hierarchy of criteria.

It appears then that we must agree with Kuhn, who claims that in choosing among theories scientists rely on certain values rather than on unequivocal methodological criteria, and that these values can conflict with one another.[1] This is a common situation occurring whenever and wherever we rely on a pluralistic set of values which are irreducible to one another, and which cannot be ordered on a single scale or into a single hierarchy.

Thirdly, the repertoire of these values is not historically constant. The demand for precision, for example, emerges only in modern science (except for astronomy), and does not appear in all disciplines at the same time. The demand for operationality appears in this repertoire only in recent times, and is also not present or observed in all fields to the same extent. I believe that it is precisely the ideals of science which co-determine the repertoires of values, although they certainly do not order them hierarchically in an unequivocal manner.

While I agree with Kuhn that these repertoires are historically changeable and can be ordered into various kinds of hierarchies, I believe that he gives too much weight to individual and narrow group preferences conditioned by specialist paradigms or disciplinary matrices, and that he does not sufficiently take into account the shared repertoires of values imposed by the accepted ideal of scientific knowledge in general, that is, by what is common to all the "disciplinary matrices."[2] Obviously, as a consequence of this view it becomes impossible to characterize the science of a given period other than by listing the various disciplines which belong to it, as was mentioned already in chapter I.

In any case, even if we take into account these additional criteria (which do not really function as criteria, and moreover often conflict with one another), their inclusion does not eliminate the methodological freedom which we discussed above. In order to understand and explain the choices made by scientists, one has first to consider the issue of the particular repertoire of values imposed by the accepted ideal of science, and secondly the preferences for specific values built into this repertoire.

The issue raised at the beginning of this chapter remains pertinent as long as it is not assumed implicitly that the development of scientific knowledge must either take place according to unchangeable methodological rules, or that it is necessary to admit that—to use the terminology of Paul Feyerabend—"science is an essentially anarchistic enterprise."[3] This (in my opinion wrongly constructed) alternative is accepted both by those who defend at any price the thesis that it is possible to reconstruct the development of science on the basis of unchangeable methodological rules (a supra-historical logic of development), and by those who—correctly denying the existence of such rules—derive from this fact the conclusion that no mechanism of development can be described at all, and that the only rule of scientific procedure is "anything goes." Viewed in terms of the manner in which theories are accepted in science, science is thus considered as not essentially different from other forms of thought, including various ideologies.

The issue raised above is fundamental in the sense that it demands a clarification of the issue of how scientists make decisions concerning the choice of theories, given that methodology alone does not and cannot offer conclusive criteria of choice.

I believe that the reason we lack an answer to this question—in any case a satisfactory answer (and what has been said above does not constitute such an answer)—is that until now philosophers have searched for it either exclusively in scientific methodology or exclusively in the sociology of knowledge. The idea that philosophy of science concerns itself only with the context of justification, while sociology, history, or psychology deal only with the context of discovery, has sanctioned this state of affairs and made it more difficult to overcome. In my opinion, in order to overcome it we do not have to deny the existence of an essential distinction between questions *quid facti* and *quid iuris*, but we have to admit that the answer to the question *quid iuris* is by no means independent of the question *quid facti*; or in other words, that the methodological *iuris* changes historically. To say that it is historically contingent, however, does not mean that it does not exist at all.

Toutes proportions gardées, philosophers today are in the position of those who had to explain how it is possible that although the earth is turning, wine does not splash from the pitcher; and not of those who had to defend the view that since wine does not splash from the pitcher, the earth cannot be turning. Extending the analogy further, let us note that there are other possible solutions to the problem: perhaps the wine does splash from the pitcher after all, or maybe the pitcher does not exist at all. In other words, perhaps it is an illusion to claim that the number of solved scientific problems has grown over time, and to believe in the existence of scientific progress. This possibility is strongly suggested by Paul Feyerabend, in his otherwise correct and convincing critique of contemporary philosophy, when he asks, "What makes modern science preferable to the science of the Aristotelians, or to the ideology of the Azande?"[4] Let us now examine Feyerabend's position more closely.

2.

Krystyna Zamiara is correct when she writes in her Introduction to the Polish edition of the works of Feyerabend that "there are thinkers who are important above all because their critique of accepted views initiates new paths for the development of scientific thought." She suggests that in

Paul Feyerabend's writings we find just such a predominance of critical over constructive elements.[5]

This collection unfortunately contains only texts written in the sixties, so that even when it was published it did not give a full account of the present position of the author and of the range of his critique, which expanded significantly following the events of 1968. The critique of philosophy of science was at that point supplemented by a critique of contemporary science as a social institution and as a form of human thinking.

Feyerabend is in my opinion the most important philosopher of science, who not only joins together these two strands of criticism, but who also bases his critique of contemporary science on a profound historical and methodological analysis of scientific development. This conjunction accounts for the exceptional importance of his writings, especially with respect to the subject of interest to us here.

As Feyerabend himself admits, the works of Karl Popper, and especially his critique of radical empiricism, were the starting point of his own philosophy.[6] In his papers of the early sixties Feyerabend extended this critique significantly, calling his own position theoretical pluralism (the name epistemological anarchism appeared only later).

Accepting (as did Popper) that the distinction between purely empirical and purely theoretical statements is an epistemological fiction, since there are no sentences which are not theory laden, Feyerabend seems to have been the first to pose the question of how it is possible to decide that one theory is better than another if the empirical statements which are to serve as the basis for such evaluation are themselves interpretations of natural phenomena in the language of accepted theories, and are thus not neutral with respect to those theories.[7]

It is difficult to underestimate the importance of this question. Without exaggeration it can be said that its consequences for empiricist philosophy are dramatic. If Feyerabend is right in this case (and I believe he is)—that is, if empirical data are always interpreted in the light of theoretical assumptions which we have previously accepted, and if by the same token identical terms have different meanings in different theories—then we have to revise two basic assumptions of empiricist philosophy of science: first, that in a given field of science new theories are acceptable only if they either subsume (as special cases) or at least are consistent with the previously accepted theories in this field; and secondly, that the meaning of (observational) terms should remain invariant during the process of scientific development, i.e., all new theories should be formulated so that when they are used to explain phenomena they will not violate the assertions of the observational reports they are attempting to explain.[8]

Both of these issues return in almost all the works of Feyerabend. He calls the first assumption the "consistency condition," and the second the "condition of meaning invariance."

Both of these conditions must be rejected for the same reason: *they give preference not to better theories but to older theories*. Acceptance of these conditions is thus also an expression of conformism with respect to existing knowledge, signifying agreement not to revise the meanings of the terms with the help of which knowledge is formulated, and requiring one to attempt to make new theories conform to old ones in this respect. It is obvious, however, that if the temporal order of the two theories were reversed, then on the basis of the same methodological postulates the relationship of dependence between the meanings of terms would also be reversed. Both of these conditions seemingly exclude the possibility of certain kinds of changes in our knowledge, changes which—precisely because of the theory-ladenness of all observational terms—in fact do take place in the development of science and which are perhaps the most interesting ones from the epistemological point of view. Thus a philosophy of science which accepts the conditions stipulated above can neither give an account of the actual mechanism of the development of knowledge nor impose a methodological *iuris* whose observance would benefit scientific progress.

In a word, Popper's falsificationism, which demands the most severe possible empirical testing of theories, is not sufficiently radical and critical. A critique of theories in the light of accepted empirical facts is not sufficient because these facts are already marked by the accepted theories which might demand revision and which are co-constituted by these facts. Both the consistency condition and the invariance condition conceal this circumstance: this is obvious in the case of radical empiricism, since it accepts the existence of an observational language which is neutral with respect to all theories. The issue is more complicated for falsificationism, since it does not recognize *de iure* the existence of such a language, although it does assume that observational reports can be formulated in a language independent from that of the theories to be tested. This assumption, however, cannot be defended for the reasons discussed in the last chapter. First, it cannot be defended because no theory, no fragment of our knowledge, can be tested in isolation; and secondly because we are unable to account for all the assumptions which are in fact implicated in the testing procedure. I believe that both of these considerations support Feyerabend's position, and they do so in a manner that is completely independent of historical examples.

If so, Feyerabend claims, then we need "a means of criticizing the accepted theory in a manner which goes *beyond* the criticism provided by a comparison of that theory 'with the facts'"[9] An alternative theory might constitute such a means, since only such an alternative theory could reveal that: (a) the hitherto accepted theory eliminated certain facts from its field of vision, or did not allow for their formulation from within its own conceptual apparatus; and (b) the facts which have been considered as confirming this theory, or could possibly be seen in this way as a result of the theoretical interpretation which they received within its conceptual framework, can, given an alternative formulation, assume an ambivalent position in this regard.

> However closely a theory seems to reflect the facts, however universal its use, and however necessary its existence seems to be to those speaking the corresponding idiom, its factual adequacy can be asserted only *after* it has been confronted with alternatives *whose invention and detailed development must therefore precede any final assertion of practical success and factual adequacy.*[10]

It is thus not the experiment alone, but theories alternative to the accepted ones—together with alternative ontological assumptions hidden behind the accepted observational language—which can and should constitute a real basis for criticism. A good empiricist must be a critical and inventive metaphysician.[11]

In cases where we do not take such alternative points of view into account, and remain satisfied when known facts correspond with accepted theory, we cannot exclude the possibility that the empirical success of a theory is more the result of our own insufficiently critical attitude (insufficiently critical in relation to facts) than of the actual logical value of the theory.[12] In this case the success of a theory, just like the success of a myth, is simply a construct of its adherents.

Let us note that in the work discussed here (as in other papers from this period) Feyerabend does not claim that science is no better than myth or religion,[13] but only that when we apply an insufficiently critical method imposed by contemporary empiricist philosophy (especially by logical empiricism), science is "on the best way to become a *dogmatic* metaphysical system."[14] Feyerabend is also not opposed to all method here, but only to methods which favor the unification of beliefs in science. He himself recommends a specific method which would be more critical, which he calls theoretical pluralism. This method "allows for a much sharper criticism of accepted ideas than does the comparison with a domain of 'facts.'"[15] He adds, "*Unanimity of opinion may be fitting for a church, for the frightened victims of some (ancient, or modern) myth, or for the weak and willing followers of some tyrant; variety of opinion is a feature necessary for objective knowledge; and a method that is compatible with a humanitarian outlook.*"[16]

In a word, Feyerabend presents himself as the critic of a bad philosophy of science in order to defend science from its consequences. He seems to believe, moreover, that theoretical pluralism is an exemplary rational research method, and that science realizes this method, or at least that it can do so. Moreover, Feyerabend did not reject empiricism completely, but rather interpreted it as a certain credible "cosmological hypothesis" concerning the relations between man and the world.[17] In his very interesting work, "Problems of Empiricism," Feyerabend successfully shows how this cosmological hypothesis itself changed its content during the process of the development of human thought, and how very different forms of empiricism have functioned in the history of science. Although radicalizing Popper's position, Feyerabend, like Popper, treats the method he proposes as a particular ethics of cognition, one which is not essentially different from that of Popper except that it provides a different answer to the question of "what to do in order to submit theories to the most severe control." If Popper demands that we always be ready to specify the empirical conditions under which we will be prepared to abandon our theoretical views, Feyerabend goes further and states that in addition, we should always keep in mind the question: "what are the theoretical assumptions on the basis of which we would be prepared to acknowledge that the 'facts' accepted thus far are not facts at all?" In this respect his position is more radical than that of Popper. His ethics of cognition demands from us not only a readiness to expose our beliefs to the severest empirical tests, but demands also that whenever a theory successfully passes such a test we must be prepared to ask ourselves the next question: what theoretical or metaphysical assumptions, different from those we accept, would require us to interpret the results of the experiments we have just conducted differently? Using epistemological categories, one could say that this ethics demands from us not only a maximally critical attitude to our own knowledge, but also a maximum of self-knowledge, in searching, controlling and criticizing all the assumptions which lead us to one rather than another evaluation of empirical results, which cease to be subject to empirical control precisely because we treat them as purely empirical.

An attitude of relativism with respect to one's own convictions is an obvious element of this ethics of cognition. It constantly enjoins us to ask what the situation would be if we held different convictions. It constantly reminds us that if we held convictions different from those we actually hold (often unknowingly and uncritically), then perhaps we would have to revise many of the beliefs which seem to us to be empirically indubitable, and that in this respect, we never know what various possibilities lie open to us. It is an ethics of insubmission to the facts.

Feyerabend knows that usually a person cannot think that he is both right and wrong at the same time, and cannot—without a split personality —specify alternative visions of the world or alternative scientific theories. An individual cannot consistently and exhaustively implement the norms of theoretical pluralism; a person will inevitably stop at some point. But while he is aware of this, Feyerabend believes that all individuals should at least strive for such a position, and that what an individual cannot accomplish alone can be accomplished by science or by the scientific community. This is why he asks about the conditions under which this could happen.

It is at this point that his critique of the scientific method changes into a critique of science as a social institution which—according to the author of *Against Method*—is unable to realize this norm. If Popper believes that science is that form of human thought which can and does rationally realize the norm of criticism, and if he, by the same token, rejects Kuhn's characterization of a "normal" science uncritical of its own paradigmatic assumptions as something normal for science (now without quotation marks), then Feyerabend appears to accept Kuhn's diagnosis in its descriptive respects, especially as regards contemporary science.[18] But while accepting this diagnosis, Feyerabend tries to vote for criticism and against any science which would be to some degree dogmatic. It is at this point that theoretical pluralism gives way to epistemological anarchism.

3.

Let us, however, return to the consequences which follow from the rejection of the conditions of consistency and invariance of meaning.

The first such consequence is the questioning of the adequacy of Hempel's and Popper's (nomologico-deductive) covering-law model of scientific explanation. This model was considered adequate at least for the natural sciences until Feyerabend questioned its relevance even there (see above, ch. II). The demand that a theory to be explained (*explanandum*) must logically follow from another theory cannot be met if the meanings of the terms occurring in both theories differ, that is, if the meaning invariance condition is not met in both theories. Thus the methodological norm demanding that scientific explanations be constructed so as to meet this condition cannot always be observed in practice, and "a formal and 'objective' account of explanation cannot be given."[19] Treating this norm as universally valid would again lead to preferences for older rather than better theories. It would eliminate explanations based on reinterpretations of the meanings of the theories being explained, and would ban the adoption of new meanings for the terms used in a theory. The same objection affects the reduction of one theory to another, since reduction is a special case of explanation.[20]

The second consequence, of special interest to us here because it directly concerns the mechanisms of scientific development and the criteria of theory choice, is the questioning of the so-called principle of correspondence between theories: the requirement that successive theories contain prior theories as special cases, that they explain everything the old theory explained, and in addition that the new theory explain the empirical anomalies which could not be explained by the previous theory. According to this principle the new theory must contain the old one and have a greater empirical content.

The principle of correspondence can be understood either descriptively, as a claim accounting for how older theories give way to new ones, or as a methodological norm indicating which theories are to be sought in situations of conflict between an accepted theory and experimental results. Not only logical empiricism and Popper's falsificationism, but also the methodology of scientific programmes have accepted this principle in both the senses indicated above, and have linked the rationality of scientific development with adherence to it. Thus the rationality of the development of science has been linked with an increase in its empirical content.

If, however, the meanings of observational terms are not identical in two theories, then Feyerabend claims that it is impossible to compare their empirical contents. A necessary condition for such a comparison is that the semantic model of the old theory be a submodel of the new one; only then can all the empirical consequences of the old theory also be consequences of the new one, which must also have some additional consequences that did not follow from its predecessor. When the meanings are not identical, the semantic models of the two theories can at best only intersect, so that the earlier theory would have empirical consequences that are not contained in the semantic model of the later theory. (In the extreme case the two semantic models do not even intersect, and are thus incommensurable.) As a result, the new theory cannot explain everything that its predecessor explained.

Such partially or entirely incommensurable theories speak two un-translatable (or not fully translatable) languages, and there is no formal criterion on the basis of which it would be possible to say that the empirical content of the new theory is richer. Invariance of meaning constitutes a necessary condition for the truth of the principle of correspondence when the principle is treated descriptively; and it is a necessary condition for the possibility of obeying this principle as a methodological norm. If this condition is not met, then we cannot say that the development of knowledge takes place according to any set of methodological rules, or that it should take place in accordance with such rules. In this case there does not exist and cannot exist any formal criterion for theory choice. This is essentially the meaning of Feyerabend's thesis that science is a "basically anarchic enterprise," and that the only rule which a scientific methodology can sensibly postulate is the principle of "anything goes." There are no methods of investigation which can be prohibited from the start, since any prohibition might slow down the development of knowledge. This is the point at which theoretical pluralism changes into epistemological anarchism with all its consequences, such as the claim that science is no better than myth, that the replacement of one theory with another incommensurable with it is a result of persuasion and propaganda rather than of rational argumentation, and so on.[21]

We will return to these consequences in more detail shortly, but first we must consider the issue of whether the development of scientific knowledge indeed contradicts the principle of correspondence. If Feyerabend is correct on this point and successive theories can be incommensurable (which does not mean that they always are), then we have to ask whether the rejection of the principle of correspondence necessarily makes spurious the concept of the growth of the empirical content of theory and implies an acceptance of the principle of "anything goes."

I believe that the first of these questions can be answered positively. The historical facts and the epistemological arguments I cited above both show that a relationship of correspondence does not always obtain between successive theories, and thus that one is not justified in demanding such correspondence from new theories. Theoretically speaking, the principle of correspondence can be defended only if one is a radical empiricist, i.e. if one believes that the empirical basis of science can be theoretically neutral.[22]

Following Feyerabend's and Kuhn's historical analyses of the semantic relations between the basic concepts of such theories as classical mechanics and Einstein's relativity theory, the Copernican and the Ptolemaic theories, or the physics of Aristotle and Galileo, it is difficult to doubt that in all of these cases—and surely they are not unique—the development of knowledge took place in violation of the principle of (semantic) correspondence.[23]

I will cite here only one of the many examples analyzed by Feyerabend, namely that of the relationship between the classical law of energy conservation and relativity theory. The principle of correspondence would require that the classical law become a special case (valid under certain limiting conditions) of the relativist law. Many authors have indeed tried to prove that such a relationship obtains whenever the ratio between the speed of a body and the speed of light is small (more precisely, when it approaches zero). It is correct that at the limit (as v/c approaches zero), the mathematical formalization of the theory of relativity turns into the mathematical formalization of classical theory, so that the formal principle of correspondence is fulfilled; but this is not equivalent to the claim that semantic correspondence obtains between these two theories. Semantic correspondence would obtain if the concepts used in both theories shared identical meanings: in this case, if the concept of "mass" meant the same thing in both the classical and the relativist law of conservation. As Feyerabend puts it:

> The first indication of a possible change of meaning may be seen in the fact that in the classical case, the mass of an aggregate of parts equals the sum of the masses of the parts: $M(\Sigma P^i) = \Sigma M(P^i)$. This is not valid in the case of relativity, where the relative velocities and potential energies contribute to the mass balance. That the relativistic concept and the classical concept of mass are very different indeed becomes clear if we also consider that the former is a *relation*, involving relative velocities, between an object and a coordinate system, whereas the latter is a *property* of the object itself and independent of its behavior in coordinate systems.[24]

Nor can the principle of correspondence be saved by identifying classical mass with relativistic rest mass. Although both can assume the same numerical value, they cannot be represented by the same concept. "The relativistic rest mass is still dependent on the coordinate system chosen (in which it is at rest and has that specific value), whereas the classical mass is not so dependent. We have to conclude, then, that [classical and relativistic mass] mean different things and that [statements of the conservation of classical and relativistic mass] are different assertions."[25]

If so, then semantic correspondence does not obtain between classical theory and relativity theory; and if Einstein had followed the normative principle of correspondence which forbids changing the meanings of terms, he could have never formulated the theory of relativity.

Other basic concepts of classical mechanics also assume different meanings within the framework of the theory of relativity. For example, inertial motion, which had been understood as the rectilinear motion of a body in an infinite three-dimensional Euclidean space, becomes motion along a geodesic in a finite but unlimited Riemann space. The use of identical terms might convey the mistaken impression that the concept of "inertial motion" means the same thing in the two theories.

But if we accept Feyerabend's claim regarding the possibility of change in the meanings of terms in successive theories, we must then ask whether the conclusions he draws from this important claim are also justified.

4.

First of all, it is not true that every new theory formulated so as to explain the anomalies faced by an old theory is necessarily semantically incommensurable with its predecessor. It can be formulated in the same language; it need not introduce any new terms, or by introducing them it need not change the meaning of the old concepts.[26] In such cases the model of the old theory does indeed become a submodel of the new one, and a relation of correspondence obtains between them. Thus, although it is true that the meanings of terms are not fixed once and for all, and that they depend on the theoretical assumptions being made, nevertheless a new theory does not always result in their being changed. It is possible to claim that at least with reference to such evolutionary changes the principle of correspondence is fulfilled.

But the argument cited above is not an argument against the position of Feyerabend, since we can never know *a priori* what sort of change will be required to remove an empirical anomaly, so we cannot claim that we should always follow the principle of semantic correspondence. And nobody has ever denied that in some cases following this principle has led to success.

However, the question arises of how and why it is that even after theoretical changes which have resulted in changes in the meanings of terms, old theories are often treated as if they were in fact limiting cases of new theories: as if the new theories basically generalized and at the same time made more precise the conditions of their validity in the old formulation. Is it not the case that although such shifts in meaning do undermine the universal validity of the principle of correspondence, a complete discontinuity in the development of knowledge does not take place even in such cases? And if such continuity is and in some sense should be maintained, then the rejection of the invariance condition does not as yet imply—either normatively or descriptively—the acceptance of the principle of "anything goes." In accepting Feyerabend's thesis that the principle of correspondence is not universally respected in the development of science, and thus that it cannot serve as a generally valid norm in the building of new theories, I nevertheless do not share his belief that this implies a complete discontinuity in the process of scientific development.

It is true that when a new theory which is supposed to solve the empirical difficulties of an old theory is being formulated for the first time, it is usually unable to deal even with all the empirical facts which its predecessor explained quite well. Acceptance of this new theory as a legitimate rival to the old one is not supported at this point by any necessary reasons; such a theory has yet to demonstrate its ability to survive, and this process takes time and sometimes great creative effort as well. I agree further that just as it is impossible to determine the point at which it is no longer rational to defend the old theory, so it is also impossible to formulate rules which would tell us how much time should be granted a new theory to demonstrate its ability to survive. The problem is to determine the criteria on the basis of which these evaluations are to be made; and again I agree with Feyerabend that such criteria cannot be unequivocal and are not historically invariant.

I believe, however, that at least one *negative* criterion always functions in science: the accomplishments of preceding theories cannot simply be thrown out, or invalidated for no reason. A new theory is not required to accept its heritage without revision, as the correspondence condition stipulates. And in this sense the development of science is not cumulative: not everything that was a fact for the old theory must remain a fact in the new one. Nor must the new theory be appraised according to the criteria of evaluation inherited from the old one, as is suggested by all views according to which the scientific method supplies us with supra-historical criteria of rationality. In this sense the development of science is not a continuous process, since it does not take place according to a permanent methodological plan of development. But a new theory must—at least in accordance with its own standards—evaluate, transform and somehow assimilate the accomplishments of its predecessors. This requirement does not dictate a particular method for building a new theory, but it does exclude certain methods.

A lack of cumulative growth is not the same as a break in all inheritance: a process which is discontinuous in one aspect can be continuous in another. Anarchistic convictions notwithstanding, no revolution, not even the most radical one, in science or in social life, can ever remake the world completely new. If this negative criterion were not respected, the history of science would be a series of completely independent episodes. From the fact that no future step is fully determined by the previous ones, it does not follow that no steps are excluded. In response to Feyerabend's metaphor claiming that one cannot demand that an expedition to the top of Mount Everest should climb the peak using the steps of classical ballet, I will claim equally metaphorically that although the climbing methods used up to now do not determine future ones unequivocally, landing on the peak from the air would not count as a climbing accomplishment. Any attempt to articulate the world with the help of a scientific theory which fails to take this negative criterion into account—even if it were possible, and even if its results were valuable—would find itself outside the scientific tradition.

Thus every new theory faces the problem of the reinterpretation and assimilation of the results achieved in a given discipline by its predecessors. Empirically verified claims formulated with the help of the conceptual apparatus of the old theory have to be reinterpreted in such a manner that they at least do not contradict the new theory. Some "facts" can be disqualified, but until they have been "disqualified" they remain facts.

This process demands in the first place a gradual change in the meanings of old terms and the acceptance of new operational rules, and acceptance of their use in empirical situations different from those in which they have been used thus far. At the end of a long chain of such slow changes, which are formally impossible to describe and often even unconscious, the original and the later meanings of terms can eventually come to appear commensurable and mutually translatable, so that only a detailed historical analysis can show how the transition actually took place. Whatever this process of reinterpretation looks like, once it has been accomplished—once the old claims have acquired new meanings and been reinterpreted in terms of the conceptual apparatus of the new theory—it starts to appear as if the relation of correspondence did in fact obtain between the two theories. *In fact, however, what corresponds to the new theory is not the old theory in its original, historical sense, but the old theory reinterpreted with the help of the conceptual apparatus of the new one.* This is why looking at the history of science through the lens of contemporary theories, as if they constituted a link in a causal chain, or as if human anatomy were really a clue to the monkey's anatomy, prevents us from understanding the abandoned theories as elements of the specific conceptual structures to which they belonged and with which they were in agreement.

In the course of such a process of reinterpretation, the semantic model of the old theory undergoes a gradual transformation and becomes something like a submodel of the new one. It can even happen that as a result of this procedure, the range of validity of the old theory turns out to be broader than was believed when it was first challenged, since as a result of the conceptual shifts caused by the adoption of the new theory, the old one can acquire the ability to explain some facts which previously appeared as empirical anomalies. After the formulation of the theory of relativity, scientists using the categories of Newtonian theory succeeded in explaining a series of events which previously had seemed to contradict this theory (like the perihelion of Mercury). It does not seem, however, that this could have been done without the prior appearance of the theory of relativity and the possibility of looking at Newton's theory through its lens.

Secondly, this process of translation does not take place without some "losses." Some sentences which made sense within the framework of the old theory cannot be meaningfully formulated at all in the new theoretical language. On the basis of the theory of relativity, for example, it is impossible to formulate meaningfully a question such as "how much time is necessary for a force F to give a mass m a velocity equal to $2c$?" Similarly, on the basis of quantum theory it makes no sense to ask about the position of a particle between two measurements.

In this manner something that was considered a "fact" within the old theory (for example, that an object has a position independent of the act of its measurement) ceases to be a fact within the new theory. Some "facts" are thus disqualified as facts. This means only that some of the earlier modes of conceptual articulation of the world have been rejected (along with some of their consequences); it does not mean that the new theory did not inherit the accomplishments of the old one, or that it should be rejected if it cannot explain everything that the old theory explained. If empirical facts do not have a purely empiricist character, if we reject a purely naturalistic understanding of a fact as something given independently of theory, and if facts are treated as a theoretical interpretation of reality, then "facts" cease to the subjects of theoretical description and become means of description. And if we are not instrumentalist in our theory of knowledge and we do not treat theories and facts *exclusively* as means of description, then we have to acknowledge that those same fragments of reality whose real existence we assume, and which we treat as the objects of our investigations, are not identical with any set of "facts" constructed through theory; these fragments can rather be represented by various sets of "facts," some of which are presumably mutually incommensurable. But then, in accepting this epistemological perspective we have systematically to *revise the concept of the empirical content of a theory*. This can no longer be defined in terms of the number of facts which this theory explains. We have to say instead that the empirical content of a theory depends on the domain of the reality to which it applies.

Questioning the thesis of the growth of empirical content in transitions from one theory to another, Feyerabend seems not to notice that on the basis of the epistemology he espouses, the concept of the empirical content of a theory must itself be reinterpreted. While correctly defending the view that new theories in science can at least sometimes change the meanings of previously used terms, he does not notice that the same stricture applies to epistemological concepts. While questioning the thesis about the growth of empirical content in the passage from one theory to another, he presents the matter as if the content of a theory depended on the number of facts it explained and as if a disqualification of certain "facts" as facts corresponded with the impoverishment of the empirical content of a theory. In short, he uses the concept of empirical content, which does not at all fit into his own epistemological perspective and which exposes him constantly to the charge that he is essentially an instrumentalist. He acts like a physicist who, instead of saying that the term "the position of a particle" cannot be applied to a particle being measured, and who is unable to say that a particle has a position independent of the act of its measurement, would claim that the particle has no position at all, or that particles have no existence at all unless they are being measured.

The concept of empirical content cannot be defined with reference to the concept of a fact, nor can it be defined quantitatively. An increase in empirical content can only mean that in revising the old articulation of the world and paying for it by disqualifying certain facts as facts, the new theory covers domains of reality which were not covered by the old theory. When we claim that the old and new theories are incommensurable, we are suggesting that we have available some means of measurement, but that we cannot apply these measurements to the objects we are comparing. In essence, the point is not that theories are incommensurable in this sense, but rather that the very concept of measurement in terms of empirical content defined in terms of facts has been destroyed. I do not believe that this concept can be replaced with another quantitative one.[27]

Thirdly, if the acceptance of a new theory is accompanied by the acceptance of new evaluative criteria, with a different hierarchy in the repertoire of values which scientific knowledge is supposed to realize, then it might turn out that certain merits of the old theory which the new theory does not share will cease to be considered as advantages. The demand that the new theory be quantitatively more precise, or that it be as general as possible, can make the theory less simple or less easy to visualize than its predecessor. Galileo's physics was at first more difficult to visualize than Aristotle's, and this was certainly considered a shortcoming. In time this shortcoming came to be seen as a strength.

Both normative and descriptive versions of the principle of correspondence can then be understood in both a weaker and a stronger sense. As a norm it can either refer to the strong requirement that a new theory *from the moment of its first formulation* correspond with the old theory, or the weaker requirement that *in time* it should at least prove able to accommodate an interpretation and assimilation such as we have discussed above. Of course not every theory will prove able to meet this requirement. As a description, the principle of correspondence can either mean that new theories are only successive stories added to an existing explanatory structure which remains unchanged as the new floors are built, or else that at least sometimes this structure is thoroughly rebuilt according to the theoretical requirements of the new order, which then "break down our objects to reconstitute them in a new [semantic] space." As a result, yesterday's planet becomes tomorrow's star. This does not mean, however, that as a result of accepting a new theory the range of validity of our knowledge does not grow, in the elementary sense that thanks to this new theory we can today interpret phenomena differently from yesterday, arrange them into new sets of "facts," and explain phenomena which we could not explain yesterday. This statement, however, demands the "metaphysical" assumption that our theories refer to a reality which they articulate, among other things, in terms of facts.

If the descriptive principle of correspondence is understood in the weaker sense, it does not follow that the process of the development of knowledge is cumulative, nor does it follow that it always takes place according to the same permanent rules which embody human rationality. What does follow from this principle is that there is no development of science without some inheritance of tradition, even if this tradition is reinterpreted and treated selectively.

If in turn the normative principle of correspondence is understood in its weaker formulation, then we are not justified in claiming that following this principle can delay the progress of knowledge. Rejection of this principle in favor of "anything goes" is tantamount to a denial of the negative criterion which was discussed above.

5.

If we return to Feyerabend's position with these points in mind, it is apparent that:

a) his epistemological and methodological analyses force us to reject only the strong formulation of the principle of correspondence;

b) the rejection of the principle of correspondence in a strong sense does not contradict the claim that a transition from one theory to another can be accompanied by growth in the empirical content of the theory, since this content cannot be defined at all in terms of the number of facts which the theory explains; the circumstance that something that was considered a fact on the basis of the old theory ceases to be a fact on the basis of the new one fails to provide any measure either for the increase or for the impoverishment of the content of our knowledge; and

c) acceptance of the rule of "anything goes" would follow only from the rejection of the principle of correspondence in its weaker sense.

As long as we accept that new theories are supposed to assimilate the accomplishments of older theories, if only through a process of reinterpretation, then by the same token we agree that not everything goes in science, since not every theoretical change leads to such a result. It is inadmissible, for example, to free theoretical propositions from the requirement that they assimilate the existing knowledge in a given area, even if they are required to do so critically and according to their own standards.

We can say that the epistemological and historical analyses presented by Feyerabend in *Problems of Empiricism* do not constitute a justification of the conclusions presented in *Against Method* and other later works. From the point of view of these analyses, the transition from theoretical pluralism to epistemological anarchism is unjustified.

I think, however, that we would be acting too quickly in declaring that we are dealing here merely with a common *non sequitur*. Feyerabend's transition does find its justification in other premises he adopts. The question is whether these additional premises are as well justified as the ones discussed thus far.

The method of theoretical pluralism defended by Feyerabend was supposed to be beneficial for the progress of science, which was understood in a specific manner. This progress was supposed to be marked by an increase in the empirical content of theories and their better correspondence with reality. Because all observational statements are theory-laden, such progress, as Feyerabend correctly noted, can be assured neither by the methodology of logical empiricism nor even by Popper's falsificationism—which was still far too liberal towards experimental data. Feyerabend's historical analyses showed that science often achieved such progress in contravention of methodological rules formulated on the basis of these views. His methodological pluralism was supposed to constitute a better means of achieving the same goal. And although, as a result of this, he viewed the development of science differently from his opponents, and claimed that this process was not rational in the way in which his opponents conceived of it as rational, he nevertheless agreed with them that increasing the empirical content of science is a value, and that science, in contrast to other forms of human thought (mythology, ideology, magic, religion), is able to realize this value. Thus at that time he did not claim that science is no better than a myth, but argued that the task is to make sure that it does not transform itself into a metaphysical dogma; and he cautioned that humility before facts by no means protects it from this fate. Questions such as "what is so great about science?" or even "what is so good about truth?" certainly did not fit within the perspective of theoretical pluralism.

A fundamental change of perspective—a transition to epistemological anarchism—finds its justification only in Feyerabend's notion that "science is not prepared to make a theoretical pluralism the foundation of research" and that science would be impossible without dogmatism.[28] Moreover, he came to believe that an increase in empirical content is not a positive value, and that the imposition of such a requirement on science (even if it were possible) would be undesirable from a humanistic point of view. When one reads Feyerabend's work from this last period it is difficult to doubt that his current position is indeed opposed to treating the growth of empirical content as a value. Thus Feyerabend argues:

> [W]e must also adopt the right attitude towards the results which modern science has produced. It has produced cars and telephones and many people cannot imagine living without them. Now, to start with, the continued production of cars and telephones does not depend on a scientific philosophy. Having invented them we can produce them by memory, and without any philosophy whatsoever. It is the constant change of cars and telephones that is in need of a philosophy of change and improvement. But most changes in these fields are not improvements, they are simply incentives to keep people buying cars. Real

improvements of life, however, may need a fundamental change of basic philosophy, and away from content increase.[29]

If I understand this statement correctly, Feyerabend is claiming that the increase of empirical content which has guided modern science as a methodological norm also serves technological progress and is connected with it as a value. At the same time, technological progress by no means serves to improve human life; on the contrary—it enslaves people, turning them into the slaves of their things (of newer and better cars, for example). Man needs a philosophy which will not treat technological progress as the highest value. A philosophy of science which evaluates the progress of knowledge in terms of the growth of its empirical content is basically just a different expression of the philosophy which treats technological progress as a superior value, and it is just as antihumanist as the latter.

From this perspective the question which Feyerabend poses repeatedly becomes clear: why should we value science more than, say, the Azande religion or Aristotelianism, which did not follow the postulate of the growth of empirical content? In one respect, Feyerabend is absolutely right here: when the question is posed in this manner, it is impossible to answer that we should value science because it leads to progress, since it is precisely the value of progress which has been called into question.[30]

I think that the worst thing that philosophy of science can do when faced with this question is to avoid answering it (as John Watkins tries to do) by claiming that "Our task [...] was to discuss criteria of scientific progress; it was not to discuss whether scientific progress is good or bad for mankind."[31] Watkins is of course welcome to discuss whatever he chooses, but philosophy of science, if it is to be a philosophy, must attempt to answer this question under the threat of self-destruction. Watkins' answer—and he is not the only one to offer such a response—is tantamount to saying that philosophy of science is interested in means but not in the ends served by these means.

First, to be interested in means and not in ends signifies in effect a silent acquiescence in these ends: after all, no one would spend his time developing means to realize ends of which he did not approve. The times are past when it was possible to believe that the approval of these ends did not require discussion, so today this answer can only be given in bad faith.

Secondly, why should anyone be interested in a philosophy of science which forgoes the task of justifying the goals it serves, especially if it claims in addition that it does not supply scientists with any heuristic directives?

Thirdly and finally, this type of answer, which in the best of cases amounts to a shrug of the shoulders when one is faced with questions about the significance of one's own activities, does not lead to a solution to the problem, but only—like any other evasion—to breaking off the discussion: "I value progress, and you do not, so we have nothing to talk about"; worse yet, it provokes the opponent into using other means—such as coercion—to try to convince. As Feyerabend puts it,

> Critical rationalists are not liable to listen to the reasons which are introduced by a wider discussion, but other people are and, when convinced, will refuse to continue being taxed for 'knowledge.' So unreasonable people will have to be educated by the financial measures reasonable people may soon take against them.[32]

> Take the *money* away from them and they will soon be reasonable.[33]

> Violence, whether political or spiritual, plays an important role in almost all forms of anarchism. Violence is *necessary* to overcome the impediments erected by a well-organized society [...] and it is *beneficial* for the individual, for it releases one's energies and makes one realize the powers at one's disposal.[34]

Let us then attempt to dispute Feyerabend in terms of this most fundamental issue which he raises.

The first and essential issue is the assumption of an unbreakable conjunction between the desire for the increase in empirical content and the pursuit of technological progress; and the conclusion drawn from this connection is that a society for which technological progress would no longer be a superior value, and which would aim instead at the "real improvement of human life," would not be interested in the growth of the empirical content of its knowledge. I think that neither this premise nor the conclusions drawn from it are correct.

This premise was addressed in chapter I, where I tried to show that the ideal of science which joins its technological and its cognitive functions, or even tries to subordinate the latter to the former, is a product of a particular culture rather than an immanent characteristic of scientific cognition, and that the criteria of rationality which it imposes are not an embodiment of an immanent rationality specific to human nature. I think that Feyerabend makes a mistake analogous to that committed by his opponents: he universalizes this ideal and treats it as supra-historical, although the conclusions he draws from this fact are diametrically opposed. While for his opponents the universalization of this ideal constitutes a basis for an unconditional acceptance of science and of its goals as the only rational ones, for Feyerabend this same premise leads to an equally unconditional rejection. Since the only possible rationality is one based on the linkage between the cognitive and the technological functions of knowledge, then —because of the negative social consequences of the realization of constant technological progress—we have to take a position against rationality. Instead of criticizing the concept of rationality imposed by this ideal of science, this concept is treated as rationality *tout court*, and it can either be uncritically accepted or equally uncritically rejected. In the first instance this leads in one way or another to the acceptance of a technocratic social order, since it is precisely this order which needs science as a tool which does not question the goals for which it is to be used, and it needs a philosophy of science which will treat these goals as unproblematic; in the second case, the rejection leads to anarchism, which identifies the existing order with the rational order and sees irrational means as the only method of transforming this order by appealing to intellectual or political terror. I think that it is precisely a supra-historical treatment of human rationality, and more generally of human nature, which places all philosophy trying to defend a liberal system of values before a rather unpleasant alternative: either a defense of the existing order or anarchism.

If, however, this junction between the cognitive and the technological functions of science is not unbreakable—if it is not a necessity of scientific reason, but a historical fact—then it raises the following question: even if one evaluates technological progress in uniformly pejorative terms, if one sees it only as a means of enslaving men (and such an evaluation is obviously rather one-sided), then on what basis can one claim that knowledge which aims at the improvement of human life should do without increases in empirical content? Regardless of the concrete content of this improvement, it seems in any case that its realization would require an increase in our knowledge of man and the world which surrounds us. In short, I can see no reason why a rejection of the view of technological progress as a superior value for civilization should necessarily have to imply an abandonment of the goal of increasing the empirical content of cognition.

One can, of course, claim that such cognition should be guided by completely different methodological norms, but I see no reason to believe that this would be a sphere of intellectual effort directed by the principle of "anything goes." One can and should claim that a philosophy of science which is blind to these alternative possibilities, and which cannot appreciate the justification of its own rationality, cannot, given the social status of science today, fulfil the functions it should fulfil: to offer a critique of the results of human activity and cognition. I see no grounds, however, for

the claim that the insufficiencies of human reason can be overcome only
if the use of reason is given up altogether. One can claim that the subordination of an entire culture to the criteria of rationality characteristic of
modern science, that is, to the criteria of technological rationality, poses a
mortal threat to cultural values, and one can demand that the pluralism of
this tradition be defended. But at the same time one should remember that
"pluralism without limits" would in consequence lead to the disintegration
of society and its culture, and to the destruction of all shared consensus
within the framework of which pluralism is possible and makes sense. At
moments, it seems that Feyerabend does notice this consequence of his
position, for example, when he claims that anarchism offers only a temporary cure for the disease of a culture which is dominated by the technological rationality of science. It seems, however, that this is only verbal
play, since he does not even ask the question of whether and when this
supposed medicine could be given in too large a dose. Even if science
were only one thread, and not the dominant one in our tradition, the question about its rationality addresses the issue of the kind of science we need
today, and not the issue of whether or not we need science at all.

PHILOSOPHY OF SCIENCE AND SOCIOLOGY OF KNOWLEDGE[*]

Cognition and Knowledge Universalization[1]

1. The Problem-Situation

Facing recent developments in the reflection on science sometimes referred to as "the sociological turn,"[2] the philosopher of science is tempted to ask: What is the epistemological significance of the ætiology of knowledge in general?[3]

By the ætiology of knowledge I mean all kinds of investigations concerning the impact of the circumstances of cognition upon its content. There can be no doubt that the sociology of knowledge and the history of science (but not only history and sociology) belong to this wide field of investigation so fashionable today.

The traditional answer to the question concerning the epistemological significance of the ætiology of knowledge given by rationalist philosophers —not only by logical empiricists or by Popperians but also, for example, by Husserlians—was decidedly negative: If Pythagoras sought the foundations of being in mathematical relations; if the Darwinian theory of evolution was born of Malthusian inspirations and Malthus' ideas sprang from liberal ideology; if Lord Kelvin in his investigations on electromagnetic theory was motivated by the utilitarian values of Victorian England;[4] if the controversy between Pasteur and Pouchet concerning spontaneous generation reflected the political controversies of the Louis Napoleon period in France;[5] and if the indeterminism of German physicists in the Weimar Republic was caused by the political and ideological atmosphere of that period in Germany;[6] nevertheless, all these determinations, even if well substantiated, should, according to these philosophers, have no impact upon epistemological evaluations, that is to say, on the acceptance or rejection of the theories and opinions in question. Psychoanalysts may claim that the theory of relativity was formulated by Einstein because of his familial complexes; or the members of the Science for People group may denounce attempts to explain social phenomena by biological considerations as the expression of a fascist or imperialist ideology;[7] but physicists or biologists should not be concerned about such circumstances when they have to evaluate the content of these theories and opinions as

* Reprinted by kind permission of the Nikolaischen Verlagsbuchhandlung and the Wissenschaftskollegs at Berlin.

true or false. Nor should philosophers. There is a difference between science and politics. When a politician says something, we are immediately tempted to question his motives or interests. But in the case of scientific claims, we ask rather whether they are true or false, well substantiated or not, no matter what may have been the motives for advancing them.

At the first glance this traditional point of view seems convincing: if the shoemaker drinks vodka, this does not mean that his products will smell of alcohol.

However, if we think about this answer more deeply, we can easily find what it presupposes: if one admits that the circumstances of cognition may have a locally selective or deforming impact, something that even a convinced rationalist would not deny, then in order to claim that these circumstances have no epistemological significance, one has to presuppose that knowledge distorted by these circumstances may be confronted with a non-distorted model. In other words, what must be assumed is the possibility of an epistemologically privileged situation, i.e., a situation in which we know that we do not have to deal with any distorting factors.

Putting the question in Cartesian terms: how can we know that the evil demon is absent, that he is not deluding us *hic et nunc*, if we know that he is present and deludes us sometimes? The Cartesian answer is well known: it is only the veracity of God that can protect us against the deceiving tricks of the demon.

Thus our first question concerning the epistemological significance of the ætiology of knowledge is whether such an epistemologically privileged situation is possible. I will discuss the problem below, in part 2.

There can be no doubt that at any given time the scientific community generally accepts certain criteria and values for apprising (accepting or refuting) scientific claims, though such criteria are difficult to codify and have always been disputed not only by philosophers but by scientists themselves. These criteria and values belong to a wider background consensus in terms of which the scientific claims advanced in a given local and specific context will be either universalized, i.e. accepted (even if not immediately) by scientists working in quite different cultural contexts, or discarded even by those who advanced and defended them previously. It is due to this consensus that the disputes and controversies occurring frequently at frontier areas of research are usually resolved relatively quickly. Since the universal character of scientific knowledge is one of its specific features, I would claim that a socio-historical analysis of the local cultural context in which certain claims were advanced cannot by itself explain why they were accepted elsewhere. What is needed in addition is an analysis of this background consensus.

However, the quasi universal acceptance of scientific knowledge in different socio-cultural circumstances, which is one of the hard facts substantiated by its development as well as its transmission in time, even if it were undisturbed, does not yet prove that a privileged epistemological situation is in fact possible in science. The answer depends on whether the background consensus is regarded as a "necessity of Reason" or as a historical fact.

If we believe that it is valid as a "necessity of Reason," that it cannot be other than it is, then indeed the content of universally accepted knowledge cannot depend on the circumstances of its acceptance, whether cultural, historical or biological; in this case the ætiology of knowledge would have no epistemological significance. Such is the main rationalist thesis concerning the evolution of science.

If, to the contrary, we believe that the consensus is valid only due to certain factual circumstances, if we do not treat it as a necessity of Reason, then the universalization of scientific knowledge is to be explained by factors co-determining this specific consensus. If the ætiology of knowledge could provide such an explanation, it would prove by the same token that genetic factors co-determine the content of knowledge even when this knowledge is universally accepted.

Let us remark, however, that in this case two different possibilities should be distinguished: if the consensus in question is supposed to be valid under all historico-cultural circumstances, then its universal validity might be explained only in terms of biological factors; I believe that this is why some scientists and philosophers are looking today in biology for a *via media* between the Scylla of epistemological absolutism and the Charybdis of relativism. This path has been chosen not only by Konrad Lorenz, Jacques Monod, Jean Piaget, Noam Chomsky and the sociobiologists, but also by Popper, no matter how important and deep may be the differences between their opinions.

If a biological explanation were admissible, then the consensus could be treated neither as a transcendental "necessity of Reason" nor as a fact relativized to historical circumstances, but as the incarnation of an historically unchanging human biological nature. A full analysis of this possibility would, however, lead me far beyond my present subject.

If, on the other hand, the background consensus is supposed to change in time, which seems to me a more plausible position, then both socio-historical and biological explanations would be conceivable, and some kind of cultural or historical relativism could not be avoided in the explanation of the development of knowledge.

Such seems to me to be the problem situation when we pose the question of the epistemological significance of the ætiology of knowledge in the most general philosophical terms.

Let me try to summarize:

a) The ætiology of knowledge would have no epistemological significance if a privileged cognitive situation were possible.

b) The ætiology of knowledge would have epistemological significance if it explained not only why some claims were advanced, but also, and first of all, why they were universally accepted. If it cannot do this, it may provide penetrating explanations of local historical particularities in the process of scientific development, but it has no epistemological significance. The local factors cannot explain the universal acceptance of scientific knowledge.

c) In order to explain the universal acceptance of scientific knowledge, the ætiology of knowledge must investigate the factors thanks to which there exists a background consensus on the basis of which scientific claims are accepted or rejected by scientific communities, and thanks to which the consensus possibly changes with time. Whether the explanation is to be provided only in biological terms, or in historico-cultural as well as biological terms, depends on the supposed historical stability of the consensus.

If contemporary sociology or the social history of science had no goals other than to explain circumstantial particularities of cognition in various social settings, there would be no problem of its epistemological significance or of its relation to the philosophy of science. Each discipline would seek to answer quite different questions: the first, to explain local particularities of cognition; and the second, to clarify the universal acceptance of some of its results in different socio-historical contexts. Let us remark that this difference is not the same as the well-known distinction between the context of discovery and the context of justification; when we ask why and how knowledge is universalized we are not obliged to exclude the question of the genesis of the criteria of its evaluation.

In fact, however, the proponents of the strong sociological program as well as the social constructivists believe that their local case-studies do have some epistemological consequences. Today we are no longer in the same situation as Kuhn when he asked: "How could the history of science fail to be a source of phenomena to which theories of knowledge may legitimately be asked to apply?"[8] The tables have been turned, and now I feel obliged to ask: Can social history and the sociology of science indeed replace philosophy of science in solving epistemological problems? If this were not what the sociological turn implies, all my remarks would miss the point.

Having presented the problem situation as I see it, I now turn to the first question: is a privileged epistemological situation indeed possible in science?

2. The Problem of the Knowing Subject

At least since the beginning of modern times, the methodological criteria for the construction and evaluation of scientific claims have been regarded by philosophers as valid *de jure*, no matter how they presented and substantiated them. On the basis of this assumption, science was regarded as the incarnation of human rationality. This assumption has found its expression in the concept of the knowing subject.[9]

According to this conception it was assumed that the knowing subject, at least as far as scientific cognition is concerned, depends neither on an inherited tradition, nor on all the accidental circumstances in which cognitive activities take place. The knowing subject was supposed to be able to overcome his physical and historical particularity and to produce knowledge that had to be accepted at all times and places, and by every other rational knowing subject. Except for particular circumstances that might perhaps distort the results of his cognitive activities, but which might be neutralized by the intersubjective control of the results obtained, such a subject was treated as if he stood completely outside the world that he investigated, as if the results of his theoretical and experimental activities depended neither on his physical make up, nor on the instruments he used, nor on his conceptual apparatus, nor on the historical situation he was living in. We could say that the philosophers endowed the human knowing subject, at least potentially, with some of the attributes of a god. The Laplacian demon could serve as a model for such a subject.

It was just this conception of the knowing subject that constituted the commonly accepted basis for philosophical discussions concerning the method which, if rigorously applied, would enable the potentially rational subject to be actually rational, to accept all and only those claims that must be accepted by everyone in all places and at all times.

From Bacon and Descartes to Carnap and Popper,[10] this concept of the autonomous knowing subject has engendered different ideas of the scientific method that was supposed to be universally valid and to express the rational abilities of human nature. The fact that for such a long time almost all philosophical reflection on science was concerned predominantly with methodological problems, was mainly due to the conviction that the scientific method is the incarnation of human rationality and the primary tool for its realization.

At the same time, this concept of the knowing subject served as the philosophical justification for postulating the autonomy of science: of its intellectual autonomy with respect to philosophy, religion or political opinions, and of its institutional autonomy with respect to churches or the state, at least since the state was becoming more and more interested in the development of science and of its applications to practical matters. Thanks to this putative autonomy, scientists could pretend to be impartial arbiters in all human conflicts which were supposed to be soluble by means of the scientific method of which they were the masters.

For a long time, the epistemological point of view according to which cognitive activity can be completely independent of the circumstances in which it takes place, and the conception of science as of an autonomous social institution, corresponded with the state of knowledge about man and to the actual social situation, in which science was not linked to the economy or to politics by any strong institutionalized bonds. This situation, regarded as corresponding to the very nature of cognitive activity, encouraged the treatment of science as if it were not a product of a definite and changeable culture that could be otherwise than it is, but as a fact of nature. It also led to the treatment of science as only a system of opinions, like, say, religion or philosophy, though it differed from them in method.

It seems obvious that as long as this conception of the knowing subject was accepted, it was impossible to concede that the ætiology of knowledge might have any epistemological significance. In this framework there was room neither for a history of science that would go beyond a chronicle of scientific achievements and failures, nor for a sociology of scientific knowledge. And, as a matter of fact, the sociology of science was born only when these opinions were undermined by developments in knowledge and by a new situation of science in the global social structure. This radical shift in the methods of doing history of science was due, I believe, to the same circumstances. As long as those opinions prevailed, however, it had to be believed that the circumstances in which knowledge is advanced may have only a distorting, or perhaps a selective, but not a constitutive impact upon its content. In such a framework neither a strong sociological program nor the conception of the social construction of knowledge was possible.

Now it seems evident to me that the conception of the knowing subject, and of the scientific method which grants complete autonomy to the content of knowledge with respect to the circumstances under which it has been advanced and accepted, has been undermined by the very development of knowledge in the course of the last hundred years.

As a result of developments in both the natural and the social sciences, the knowing subject can no longer be treated as a subject dwelling outside the world he is investigating. On the contrary, his cognitive possibilities have been more and more relativized with respect to that world and his relations to it. The autonomy of the knowing subject, his ability to achieve knowledge unmediated by his own natural and social constitution, is questioned by physics, biology, and neurophysiology, as well as by linguistics, cultural anthropology, sociology and history (including history of science), not to mention philosophy. The great achievements of contemporary science—Einstein's theory of relativity, Heisenberg's principle of indeterminacy, the Gödel theorems—all seem to show that the more we know about the world, about ourselves and about how we know, the more difficult it is to believe that our knowledge does not depend at all on our own biological make-up, the functioning of our brains, the language we use, the culture we inherit, or the social situation in which we live.

Husserl was well aware of these philosophical consequences of the development of knowledge, and of the danger involved in the relativization of all the values of our culture. His whole intellectual effort was directed towards overcoming this danger, by finding metaphysical foundations to uphold the universal validity of our knowledge and values no matter what the circumstances of our life. It seems that he did not succeed in this global enterprise.

At the other end of the philosophical spectrum, the conception of a pure empirical basis on which all scientific knowledge is, or ought to be, based, the conception advocated by logical empiricism, was supposed, at least in the first period of its evolution, to accomplish the same task as the Husserlian conception of the transcendental ego. It was supposed to grant scientific knowledge independence from any and all circumstances under which it is achieved and accepted. This effort did not succeed either, though obviously for quite different reasons.

More recently the same philosophical aim has been pursued by Popper in his epistemology without the knowing subject, known also as the theory of "world three."[11] Contrary to what he said in *The Logic of Scientific Discovery*, Popper now agrees that the knowledge advanced by the knowing subject can never be altogether objective, free from all circumstantial and non-rational co-determinations. The concept of the rational method which, if applied, was to guarantee the objectivity of the results of human cognitive activities, is now interpreted as a kind of impersonal mechanism according to which science develops in the Platonic world of ideas and problems, and this mechanism is seen as an extension of natural selection. I would argue that as a result of the developments of scientific knowledge, as well as of the history of science, Popper faced an alternative: either he had to abandon the idea of a historically permanent rationality of scientific knowledge based on the assumption of the autonomy of the knowing subject, or he had to abandon the knowing subject altogether and move into "world three" where, due to its impersonal character, no subjective or circumstantial, non-rational factors could have any impact upon the universalization of the knowledge produced in world two. The reason why he chose the second option seems evident with respect to the main tenets of his philosophy and the role he expected that science can and should perform in our culture.

To summarize: if we can believe neither in a Cartesian God who protects us against the tricks of the evil demon, nor in transcendental reduction, nor in a pure, epistemologically unquestionable empirical basis of scientific knowledge, nor in an autonomous mechanism of the evolution of the world of ideas, we cannot avoid the conclusion that the ætiology of knowledge has epistemological significance.

As far as philosophical realism is concerned, it seems rather unrealistic to believe that one day we shall be able to look into a well so deep that we will not see our own face at the bottom. In other words, everything we know, we know as humans: no super-human point of view is possible; the content of our scientific knowledge is determined by the object under study as well as by other factors whose impact is constitutive and should not be disregarded by epistemology. A privileged epistemological situation does not exist, although obviously not all epistemological situations are equally good, not everything is possible, the object under study imposes its constraints and frustrates some human designs. The Kantian conception of *a priori* knowledge may well be essentially correct, providing "the a priori" is not transcendental but determined by genetic (biological, historical, socio-cultural) factors.

3. Does Science Exist At All?

Thus far, I suppose there are no essential disagreements between the philosophical points of view I have presented and the general assumptions of sociology of science, except perhaps for my opinion concerning the epistemological significance of investigations of the circumstances under which scientific claims are advanced. But one more remark is needed.

I have chosen to speak about the ætiology of knowledge in order to pose the problem in the most general terms. However, since the ætiology of knowledge embraces different kinds of investigations of the impact of the circumstances of cognition upon the content of knowledge, we should remember that to affirm that the ætiology of knowledge has epistemological significance does not imply *a priori* that all the co-determinations we have to look for are to be explained in terms of social structure or interests, at least directly. The history of ideas, for example, is also a part of the ætiology of scientific knowledge. Thus, we should not exclude the possibility that the impact of social factors may be mediated by ideas and values commonly accepted in the scientific community at a given time. Such a conception might contribute to the explanation of the process of the universalization of knowledge produced in different socio-cultural contexts.

Two years ago at the colloquium on Alexandre Koyré held in Paris, Yehuda Elkana read an interesting paper presenting Koyré as a "sociologist of disembodied ideas." Personally, Koyré was rather skeptical—to say the least—about the sociological approach to the evolution of knowledge, and it seems that he would not have been very happy with Elkana's description. In fact, however, what Koyré achieved in the history of science might indeed have led to sociological questions concerning the quasi-universal acceptance of "a framework of ideas within which science progresses [...] a framework of fundamental principles and axiomatic evidence which has traditionally been considered as properly belonging to philosophy."[12]

Elkana called this conceptual framework "the image of science." In the present volume I have called it "the socially accepted ideal of science," and I have tried to explain how such ideals, constituting the historically changing background consensus within which scientific activity takes place, can impose on it certain commonly accepted values, methodological rules of theory construction and explanation, and certain criteria of rationality, and can by the same token explain the universalization of scientific knowledge. Accordingly, I would regard the actual history of science as the realization of a series of different, consecutive and competing, socially accepted ideals of knowledge.

Without investigating and explaining the existence of such a background consensus (which obviously may be somewhat differently articulated in different disciplines and even in the work of different scientists), the sociology of knowledge cannot proceed beyond the study of specific case-studies of knowledge production. And what is more, in presenting these case-studies sociologists of knowledge are often tempted to treat the content of knowledge as an unmediated result of the local circumstances in which cognition takes place.

I do not know whether the program I am suggesting would be deemed "strong enough" by the sociologists; but I believe that it offers the only way to give an account of the evident specificity of science with respect to other products of human intellectual activities if we cannot accept the idea of the supra-historical rationality of human nature.

What differentiates such a program from the old so-called "rationalist tradition" is the thesis that the background consensus is not the incarnation of immanent human rationality, and that it is not historically stable. What differentiates this program from (at least some) contemporary developments in the sociology of science is the notion that if the circumstances of cognition have any impact upon the content of knowledge, this impact is not immediate, but rather is mediated by the relatively stable set of values and ideas constituting the research tradition. It is precisely on the basis of these traditions, which provide the resources for creative renewal from within, that new scientific knowledge is universalized.

Thus, the first point of my disagreement with current developments in social history and the sociology of knowledge is the fact that its proponents fail to ask the question which seems to me fundamental: namely, how does knowledge achieved under specific circumstances become universalized? Most of them concentrate on the detailed study of the impact of more or less local circumstances of cognition which cannot explain the universalization of scientific knowledge. Universalization is a specific feature of science if we compare it with all other systems of beliefs or opinions—philosophy, religion, morals, arts, customs and so on. Yet when sociologists of knowledge do speak about universalization, they usually discuss it only in terms of the repeatability of experiments by means of commonly used instruments; while the universal acceptance of theories, which is obviously a different matter, is usually ignored.[13]

Now I believe that there are two main reasons for this lack of interest in investigating the background consensus within the framework of which science is practiced at any given time.

The first reason, it seems to me, is that most sociologists of science simply do not believe in the existence of such a common background consensus in science. This is the case not only for extreme social constructivists like Latour and Woolgar, who claim that external "reality cannot be seen to have any discernible effect on the results of investigation which are manufactured and whose solidity is only a social construction."[14] It is also the case for more cautious authors who go beyond the "ethnomethodology of laboratory life" in Latour and Woolgar's sense and argue for the need to investigate the broader context of the "political economy of practices" or the "ecology of practices."[15] But even in this case, the question of the background consensus shared either by the disciplinary community of specialists or the scientific community as a whole is never raised.

According to this view, science is a set of differently oriented practices, but there is no science as a culturally determined whole, and there are even no individual sciences. There is no physics, but only specialized fields of research; and the extent of such specialization is decided "empirically," on the basis of the actual institutionalization of research: what are the university departments, research institutes, journals, instruments used, or research problems in which certain groups of scientists are involved? and so on. There is no community of physicists, but only communities of people who share certain common practices. So, for example, Timothy Lenoir writes:

> Physicists [...] do not appear as homogeneous group with a unified culture, but as subcommunities with different knowledge, constitutive interests and with different experimental traditions organized socially in terms of access to different resources and oriented around different repertoires of techniques and apparatus.[16]

What the term "practices" means is not easy to understand. Sometimes, as in the Marxist tradition, practices are evidently opposed to theoretical activities and are called "technical practices" (the question of whether such practices are free from theoretical components and whether the opposition is sound must be left open here). Sometimes the term is used in the broader sense in which any human activity is a practice, and the terms "theoretical practices" or "interpretative practices"[17] are introduced in the sense that Althusser, Foucault or Bourdieu used them. But do these terms have a meaning that is different from the terms "interpreting" and "theorizing"? If everything that man does is a practice, why should the term be used in such an equivocal way? I suspect that it is being used as a persuasive device: when science is regarded as a set of different practices, it is much more convincing to speak about its direct social determinations, since it is commonly believed that human practical actions are determined or motivated by conscious or unconscious interests or by social circumstances. Where thinking is concerned (and particularly where as systems of statements are at issue), this belief is not so universally accepted. Be that as it may, the term surely tends to blur the differences between various kinds of human activity. And at the same time—due to its common meaning—it reinforces the idea that science is rather a way of doing, of producing something (a telegraph, a bomb, a laser, a drug) than of knowing something in an abstract way. I will return to this point in the final part of this text.

This conceptual disaggregation of science into practices is perhaps one of the side-effects of the Kuhnian program. It was precisely Thomas Kuhn who, by the stress he put on the study of narrow communities of specialists sharing the same paradigm, opened the way for the conceptual disaggregation of science not only into the sciences but further into narrow specialties and finally, as it turned out, into a set of unconnected practices which have to be assembled.

Kuhn, for example, never spoke, as Koyré did, about revolutions in science consisting in the change of the "cadre des idées" in which science progresses, but rather about revolutions in paradigm-oriented narrow specialties. This is one of the main differences between his perspective on the history of science and that of Koyré which I cited earlier. And in both frameworks we are tempted to ask certain definite questions and disregard others. For example, in Kuhn's framework there is no room for investigation of a background consensus wider than that of different disciplinary paradigms or of disconnected "language games." By the same token, there is no room for philosophical questions about the scientific enterprise as a whole. (Kuhn himself did not go quite as far as his continuators, but he certainly blazed the trail.) However, if there is no science but only specialties or practices (whatever these last terms may mean), there can be no philosophy of science and no genuine philosophical problems concerning the whole enterprise. Both are reduced to more or less local problems of the social construction of knowledge and the methodology of practices. As a consequence, the question of the difference between science and other human intellectual and practical activities must be regarded as obsolete. The way is open to ask the famous question: "What is so great about Science?" even if it is not explicitly stated that science is no better than Azande mythology.

Does this conceptual disaggregation of science correspond to reality? Are the terms "science" and "scientific community" today just names for a set of disconnected activities or different narrow communities of specialists having nothing in common with one other—neither a common method, nor traditions, nor criteria for evaluation of their results, nor aims and commonly shared values? Do they play no common cultural role in human life—no matter how they are evaluated? I do not think that the actual process of specialization in science has gone as far as this conceptual disintegration presupposes.

Surely we no longer believe in conclusive criteria of demarcation, but the fact that we are not able to draw such a sharp demarcation line between different activities does not mean that there are no essential differences between them at all.

No doubt scientific activity is no longer exclusively a disinterested search for truth (and maybe it never was), and it produces not only systems of statements but also utilities. This does not mean, however, that the disinterested search for truth is not a value in scientific activity, and that science is not a system of theoretical statements. And scientists usually have ways of deciding what is true and what is not other than by negotiation alone, unless the term negotiation simply means debate. I agree with Peter Galison when, in polemics with extremists like Latour, he says that experimentation should not "be parodied as if it were not more grounded in reason than negotiations over the price of a street fair antique."[18] The stabilization of prices on the market, even on the free market, is not the best metaphor for the way in which the results of scientific activity are universalized. There is a difference between negotiation or bargaining and scientific discussion.

No doubt, again, that the contemporary scientific community is not the "république des savants" that the enlightenment philosophers dreamed of, but this does not mean that there is no community at all. There are reasons why not all human activities are regarded as scientific; even the most radical sociologists of science do not choose the objects of their investigations arbitrarily: they study certain specific practices and they do not investigate others. This means that in choosing their case studies they share some idea of what science was and what it is. What is this idea? What is the image of science or the scientific ideal they share?

4. The Anti-Theoretical Turn

The second reason for the lack of interest in investigating the long-term background consensus lies, as I believe, in a radical opposition to the so-called "theory dominated" approach to science represented by the traditional history of ideas in contrast to social history. The point is that if such a consensus exists, it is obviously a consensus of ideas and values commonly shared and transmitted in the scientific community. And the programme of explaining the development of science by certain commonly accepted ideas, even if these ideas are supposed in the final analysis to be socially co-determined and historically contingent, is regarded as not strong enough.[19]

This opposition to the theory-dominated approach has its origins both in a particular vision of science and in certain epistemological presuppositions.

According to this vision, science, and especially contemporary science, should be treated as a set of skills for producing and controlling phenomena rather than as an abstract system of statements about the world expressed in the form of general laws and theories suitable for applications in different domains of human activities. When Ian Hacking says "think about practice not theory," and states that "engineering not theorizing is the best proof of scientific realism," and when he cites Marx's famous dictum that "the point is not to understand the world but to change it"[20] (Marx was talking not about science but about philosophy), he seems to express precisely this ideal. And this is the reason why his book was so welcomed by the social constructivists.

No historian of science, to my knowledge, has expressed this vision of science so forcefully and explicitly as Norton Wise in his interesting study of William Thomson, especially in the essay entitled "Mediating Machines":

> The theory-dominated approach [...] divorces our knowledge from what we do; it also divorces it from what we care about, from our purposes, especially from our societal and political purposes. Thus, for example, it separates pure science, whose reference is *supposedly* to nature, from applied science, whose reference is to our purposes.[21]

And he explains:

> More generally, when we conceive nature itself as the source and referent of our knowledge, we deny the essential relevance to our knowledge of any so-called external factors.[22]

If I understand correctly, this means that what science is about is not nature but our practices of producing and controlling phenomena, and that if we regard nature as the source and referent of our knowledge, no sociology of science is possible. The last opinion would amount to saying that only a sociology of practice and not of abstract knowledge is possible, and that the methodology of science should spring from the "methodology of practices." If this opinion were correct, the idea of asking about the sociocultural determination of the background consensus would be nonsensical.

I suppose that after what I have said above I will be not regarded as a defender of philosophical realism. It is, however, one thing to say that our knowledge about nature is mediated by different factors (biological, cultural, social) and for that reason cannot represent nature "as it is," independently of our cognitive activities, so that no epistemologically privileged cognitive situation is possible, and quite another thing to state that theories are not about nature at all, but about practices of producing and controlling phenomena. Until now astronomers, cosmologists, geologists, anthropologists and linguists have neither produced nor controlled phenomena; what then are their theories about? The claim that our knowledge refers to nature does not imply that our theories are exact images of the world.

On the other hand, though much scientific knowledge is of course produced for application, "the most highly valued knowledge is produced for the consumption and use of colleagues in the process of producing innovations themselves."[23] This is one of the important reasons why the model of applied science, even if it were adequate, cannot be used as a model of all science.

"Long lived theoretical entities which don't end up being manipulated, commonly turn out to have been wonderful mistakes,"[24] says Hacking. I do not know how long is long enough to make this judgment true, but it took more than two thousand years before we could manipulate atoms. When Lysenko denounced genetics, one of his arguments was that we cannot manipulate genes, that they are "metaphysical entities." Generally speaking: this is not the first time that taking the state of one discipline for a model of all science turns out to be a "wonderful mistake." The ideal of intervening, as opposed to that of representing, is the ideal of a culture in which the possibility of manipulating things is regarded as the supreme value. Accordingly science is regarded as a means of production, and the process of cognition is treated as production or manufacturing. Even the language used by social constructivists is adequate to this vision of science.

I have no doubt that today this ideal shapes not only some trends in the contemporary reflection on science but also an important part of the scientific enterprise itself. But this is not a sufficient reason for accepting it. As Henri Poincaré said, "La science a sa cuisine mais elle n'est pas qu'une *cuisine*."[25] I think he was right.

Thus the ideal of science we are invited to accept is that of Bacon rather than that of Descartes or Galileo, that of Thomson rather than that of Maxwell, not to mention Einstein. It is an image of applied sciences radically opposed to that of pure theoretical sciences, an ideal of intervening, not of representing. Theories are appreciated in so far as they help practices; they are regarded almost as their emanations constructed in definite local social contexts.

"The truth of science should be located in the stable assemblies of practices,"[26] not in theories, according to Norton Wise, and he explains: "Bacon's forceful dichotomy between deductive and inductive strategies applies if we read inductions as the assembly of practices."[27] And I believe he is right: what the social constructivists are defending is a kind of inductivism.

We must still ask, however, whether this inductivist model is adequate at least for the applied sciences.

The fact that Thomson's inductive-utilitarian methodology was defeated in competition with the Maxwellians is seen by Wise as "a social victory of the deductive-theoretical ideal of physics over the practical."[28] In my opinion this "social fact" (and what fact is not social?) resulted from the development of scientific knowledge, which since the latter half of the nineteenth century has made technological process increasingly dependent upon the construction and application of abstract theories, and not on inductive improvements by technicians, even if they were theoretically-minded technicians. In other words, I believe that Thomson was defeated by the Maxwellians because the development of science, and especially of physics, made his Baconian-inductivist ideal of science obsolete. (The question of why Thomson defended this ideal is quite a different matter, and the answer proposed by Wise seems convincing.) This victory was the outcome of a long historical process which led to the development of the modern applied sciences, which have been based increasingly on the application of abstract theories.

So I would say that the anti-theory dominated approach to the history of science runs counter to the pattern of the development of science, including that of the applied sciences. Their evolution was due, as Koyré said, to the imposition on *techne* of rules of exactness which were hitherto specific only to the *episteme*.[29]

The inductivist approach has one further consequence in the making of the social history of science, namely the belief that case-studies concerning the production of knowledge in specific local circumstances may provide knowledge not only about these particular cases but an adequate vision of science development in general. It seems, however, that I have said enough to relieve me of the need to explain why this assumption seems to be an inductivist illusion.

5. Concluding Remarks

It seems to me that contemporary developments in the sociology of science accept an oversocialized conception of man. Not only is everything in man regarded as social, but what is more, sociological explanations are seen as the only valid explanations of all aspects of human life. I would say that this is a new kind of reductionism—a sociological reductionism. This is why any other program of explaining the development of science is said to be not "strong" enough.

I agree with Andrew Lugg's observation that the meaning of the term "social" is not exactly the same when it is said that man is a social animal, and when it is postulated that all his activities and their products should be explained in sociological terms.[30] Science is a social activity in the sense that it is a collective activity, but this does not mean that its content is the immediate expression of a local social structure or of particular group interests.

But even setting this equivocation aside, it seems to me that the fact that today we cannot defend the old conception of the cognitive subject does not mean that all cognition is directly determined by social factors. The oversocialized conception of man fails to take into account certain fundamental facts of human life: some human activities and abilities, even if in the final analysis they prove to be socially induced, may reach such a high level of autonomy with respect to their "final" causes or sources that their sociological explanation may be misleading. Sociological reductionism, like other forms of reductionism, underestimates this fact. We underestimate something important in human affairs when we explain them in terms of their final social determinations—not because these determinations are fictions, but because they are not strong enough to explain all their far-mediated results. It is not true that we are autonomous knowing subjects; but it is also not true that we cannot overcome our social determinations at all, in any sphere of intellectual activity. Thus, men are capable of disinterested action and of a disinterested search for truth. The explanation which asserts that a disinterested action must be motivated by some hidden social interests amounts to the claim that any disfunction in human behavior is functional. It seems to me to be a poor explanation. And if everything is determined by interests, the term loses all its cognitive importance. Its only function becomes the propagation of an assumed conception of human nature. And if we accepted this assumption, we would be authorized to ask the question: What interests determine the attempt to explain everything in terms of interests?

I refuse to ask and even more to answer this question, since I do not accept the presupposition. I do not accept this presupposition either as a universally valid statement of fact about human nature, or as a universally valid methodological norm. In the first sense it seems to be false, while in the second it is meant to explain everything. And any rule that is meant to explain everything in fact explains nothing.

Let me be so naïve as to believe that at least sometimes when we believe something, and when we express what we believe, and act accordingly, we are motivated not by interests, but by the search for truth (even if we do not agree about its definition and criteria). Or to put it differently: it is in our common interest to find the truth. However naïve, this belief seems to be a necessary precondition for defending our culture from deliberate manipulations that endanger its survival in the world we are living in.

Postscript

As usually happens in such discussions, none of us was completely happy with the way his point of view was understood and presented by his opponents. In this short postscript I would like, however, to avoid polemics and rectifications in these matters as far as possible. Our texts and the works to which we referred are available to the reader, who will be able to decide for himself which arguments seem convincing and which miss the point. The main controversy was explicitly stated, and it makes no sense to repeat the same arguments once more. As long as the problem of the universalization of scientific knowledge remains unsolved, social constructivism cannot pretend to provide an explanation of the general pattern of the development of scientific knowledge.

There is at least one point, however, about which I feel I was not expressing myself clearly enough to avoid misinterpretation, namely in my final remarks regarding the oversocialized conception of man, where I renounced a sociological reductionism which treats all human actions, intellectual as well as practical, as immediately determined by social situations and interests. The conception of the knowing subject as an instance unable to resist social or cultural pressures at all seems to me no less simplistic than the conception of the fully rational, autonomous subject I spoke of in part 2 of my presentation. In this sense a disinterested search for truth must be regarded as possible, if only, for methodological reasons, we do not accept the argument that the resistance to some pressures necessarily means submission to other and stronger constraints. I have explained why this argument seems unacceptable to me.

Two arguments were advanced against this point of view: one to the effect that there is no necessary "contradiction between the pursuit of interests and the pursuit of truth," and the second stating that "the disinterested pursuit of truth is a dangerously naïve myth, dangerous because it provides a standard argument today for researchers to work on whatever project they wish without responsibility for the interests that support it. [...] If we wish to avoid political manipulation we would be far better off to examine with our eyes open the conditions of interested activity under which knowledge is regularly produced."[31]

As far as the first argument is concerned I would answer that it would be sound if and only if it spoke about universal and not particular group interests as sociologists usually understand the term. If the "interests" in question are the interests of particular groups, and the pursuit of truth is immediately determined by such group interests, then in order to say that "there is no contradiction between the pursuit of interests and the pursuit of truth" we must accept the thesis that there exists a group whose interests are universal human interests—be it the working class, as Marx claimed, or the intellectuals, as Mannheim claimed. It seems that none of us accepts this conception.

As far as the second argument is concerned, I think that if disinterested cognition were not possible, we could not expect that somebody might, against his interests, refuse to participate in research that should be condemned for moral or political reasons. Such participation can be morally or politically condemned only if disinterested action is possible in principle, if we are not utterly determined in all our actions by the social situations within which we live.

What I said obviously does not mean that all scientific research is motivated by a disinterested search for truth. It means only that such a search for truth is possible and that it is dangerous for our culture to deny this possibility. The oversocialized conception of man reinforces this danger.

Understood in this manner, my thesis cannot be used as an "argument for researchers to work on whatever project they wish without responsibility for the interests that support it." This would be a misuse since in order to use it in this way, it must be presupposed that the search for truth is the highest value and should never be subordinated to other values. If we do not accept this presupposition—and none of us, it seems, does accept it—then we agree that in some social situations the disinterested search for truth should in some areas be postponed for better times.

NOTES AND REFERENCES

Preface to the English Edition

1. Roma: Edizione Theoria, 1986.
2. *The Structure of Scientific Revolutions* (Chicago: University of Chicago Press, 1970), 8.

Introduction

1. Carl Friedrich Weizsäcker, *Die Einheit der Natur* (München: Carl Hansen, 1971).
2. Herbert Marcuse, *One-Dimensional Man* (Boston: Beacon Press, 1964), 166.
3. Jacques Ellul, *The Technological Society* (New York: Vintage, 1964), 434.
4. Paul K. Feyerabend, "The Methodology of Scientific Research Programmes," in his *Problems of Empiricism. Philosophical Papers*, vol. 2 (Cambridge: Cambridge University Press, 1981), 203.
5. Krzysztof Pomian, "Działanie i sumienie," *Studia Filozoficzne*, 1967 (3), 65.
6. Yehuda Elkana, "Cultural Systems and Science," in R.S. Cohen, P.K. Feyerabend, and M.W. Wartofsky (eds.), *Essays in Memory of Imre Lakatos*, Boston Studies in Philosophy of Science, 39 (Dordrecht: Reidel, 1976); and his *The Problem of Knowledge in Historical Perspective* (Athens, 1973).
7. Paul K. Feyerabend, "Problems of Empiricism, Part II" in Robert G. Colodny (ed.), *The Nature and Function of Scientific Theories*, (Pittsburgh, University of Pittsburgh Press, 1970), also his "In Defence of Aristotle: Comments on the Condition of Content Increase" in G. Radnitzky and G. Andersson (eds.), *Progress and Rationality in Science*, Boston Studies in Philosophy of Science, 58 (Dordrecht: Reidel, 1978) and *Science in a Free Society* (New York: NLB, 1978).
8. Paul K. Feyerabend, *Against Method* (New York: NLB, 1975); and Hilary Rose and Steven Rose, "The Incorporation of Science," 1974 (mimeographed text).

Chapter I. The Development of Knowledge and the Ideals of Science

1. Florian Znaniecki, *The Social Role of the Man of Knowledge* (New York: Columbia University Press, 1940).
2. Krzysztof Pomian, "Filosofia-Filosofie," in *Enciclopedia Einaudi*, vol. 6 (Torino: Einaudi, 1979).
3. In my book *Between Experience and Metaphysics* (Dordrecht: Reidel, 1975), while basically supporting the view expressed here, I was not sufficiently consistent. I claimed that science begins where there appears a demand that practical and cosmological knowledge—*techne* and *episteme*—be combined into a rational, logically coherent whole. I suggested at the same time that this formulation is something more than a terminological convention, and I did not notice its evaluative character. Today I think that the critics who have noted this inconsistency are right, and that my formulation was also dictated by the acceptance of a certain

ideal of science. Regardless of whether or not one accepts this ideal, it is an illusion to treat a definition which is based on it as axiologically neutral.

4. Leszek Kołakowski, *Kultura i fetysze* (Warsaw: PWN, 1967), 191-192.

5. Herbert Butterfield, *The Origins of Modern Science, 1300-1800* (London: Bell & Hyman, 1949), 180.

6. Thomas S. Kuhn, *The Structure of Scientific Revolutions*, (Chicago: University of Chicago Press, 1962); see also Postscript to the 2nd edition, (Chicago: University of Chicago Press, 1970).

7. Paul K. Feyerabend, "Consolations for the Specialist," in his *Problems of Empiricism. Philosophical Papers*, Vol. 2 (Cambridge: Cambridge University Press, 1981).

8. Amsterdamski, *Between Experience and Metaphysics*.

9. Thomas S. Kuhn, *The Essential Tension* (Chicago: University of Chicago Press, 1977).

10. Paul K. Feyerabend, *Against Method* (London: NLB, 1975).

11. Karl R. Popper, *Objective Knowledge* (Oxford: Clarendon Press, 1972) and "Normal Science and Its Dangers," in I. Lakatos and A. Musgrave (eds.), *Criticism and the Growth of Knowledge* (Cambridge: Cambridge University Press, 1970).

12. Imre Lakatos, "History of Science and Its Rational Reconstructions," in *The Methodology of Scientific Research Programmes. Philosophical Papers*, Vol. 1, Eds., John Worrall and Gregory Currie (Cambridge: Cambridge University Press, 1978).

13. Stanisław Ossowski, "Nauka o nauce," in *O Nauce, Dzieła*, vol. 4, (Warsaw: PWN, 1967), 102.

14. It is often said of Greek science, from the perspective of the modern ideal of science, that it did not separate itself from philosophy and did not achieve maturity and independence. But from the point of view od the ideal of cognition accepted at that time, such a distinction would be nonsensical: the knowledge thus separating itself would cease to be *episteme* and become either *techne* or *doxa*. This view assumes additionally that such a separation was a "necessity of reason."

15. Alexandre Koyré, *Études d'histoire de la pensée philosophique* (Paris: Colin, 1961), 312-13.

16. Krzysztof Pomian, "La rationalité, irrationalité et la science," *Annales: Economie, Société, Civilisations* 30 (5), 1132.

17. Karl Popper, *The Logic of Scientific Discovery* (New York: Harper and Row, 1959), 52-53.

18. *Ibid.*, 55.

Chapter II. Ideals of Science and Rules of Explanation

1. Karl Popper. *Objective Knowledge: An Evolutionary Approach* (Oxford: Clarendon Press, 1972), 191.

2. In addition to the other works of Hempel and Popper mentioned in the Bibliography, see also C.G. Hempel and P. Oppenheim, "Studies in the Logic of Explanation," *Philosophy of Science*, 15 (1948); R. Braithwaite, *Scientific Expla-*

nation (Cambridge, Cambridge University Press: 1955); E. Nagel, *The Structure of Science* (New York: Harcourt, Brace and World, 1961).

3. Stefan Amsterdamski, "Ripetizione" in *Enciclopedia Einaudi*, vol. 12 (Turin: 1981), 76-86.

4. Michael Scriven, "Explanations, Predictions and Laws," in H. Feigl and G. Maxwell (eds.), *Scientific Explanation, Space and Time*, Minnesota Studies in the Philosophy of Science, 3 (Minneapolis: Minnesota University Press, 1962), 196.

5. *Ibid.*, 192-193.

6. *Ibid.*, 175-76.

7. *Ibid.*, 205.

8. *Ibid.*, 172.

9. *Ibid.*, 192-93.

10. *Ibid.*, 198-199.

11. E.E. Evans-Pritchard, *Social Anthropology* (London: Cohen & West, 1951), 57.

12. Alan Donagan, "The Popper-Hempel Theory Reconsidered" in W.H. Dray, ed., *Philosophical Analysis and History* (Westport, CT: Greenwood Press, 1966), 146.

13. Stefan Amsterdamski, "Lege," *Enciclopedia Einaudi*, vol. 8 (Torino: Einaudi, 1979), 83-129.

14. Herbert Butterfield, *History and Human Relations* (London: Collins, 1951), 146.

15. Carl Hempel, "Explanation in Science and History," in R.G. Colodny, (ed.), *Frontiers of Science and Philosophy* (Pittsburgh: University of Pittsburgh Press, 1962); Ernest Nagel, *The Structure of Science* (New York: Harcourt, Brace and World, 1961); Karl R. Popper, *Objective Knowledge*.

16. Edgar Zilsel, "Physics and the Problem of Historico-sociological Laws," *Philosophy of Science* 8 (1941), 577.

17. Paul Ricoeur, in Claude Lévi-Strauss, "Réponses à quelques questions," *Esprit* 31 (11 Nov. 1963), 640.

18. Lévi-Strauss, *Ibid.*

19. Edmund Mokrzycki analyzes the consequences of the acceptance of methodological models developed by philosophers of science for the natural sciences as valid also for the social sciences. Many of his critiques appear to me to be correct and convincing. However, he does not notice—or at least does not point out explicitly —that the debate in which he is engaged is methodologically unresolvable, and that his own position is conditioned by an ideal of science (sociology) different from that held by his opponents. See his *Philosophy of Science and Sociology* (London: Routledge and Kegan Paul, 1983).

20. Paul Feyerabend,"Explanation, Reduction, and Empiricism," in H. Feigl and G. Maxwell (eds.), *Scientific Explanation, Space and Time*, Minnesota Studies in the Philosophy of Science, 3 (Minneapolis: Minnesota University Press, 1962), 28-97.

Chapter III. The Modern Ideal of Science

1. Descartes, *Discourse on the Method of Rightly Conducting the Reason* in *The Philosophical Works of Descartes*, trans. E. Haldane and G.R.T. Ross (Cambridge, Cambridge University Press, 1972), 119.

2. Stefan Amsterdamski, "Previsione e possibilità," *Enciclopedia Einaudi*, vol. 8 (Torino: Einaudi, 1979), 83-129.

3. Herbert Butterfield, *The Origins of Modern Science* (London: Bell and Hyman, 1958), 179.

4. Paul K. Feyerabend, "Problems of Empiricism, Part I, in R.G. Colodny, (ed.), *Beyond the Edge of Certainty: Essays in Contemporary Science and Philosophy* (Englewood Cliffs, N.J.: Prentice-Hall, 1965), 154.

5. Alexandre Koyré, *Études d'histoire de la pensée scientifique* (Paris: P.U.F., 1966), 152.

6. "With Galileo experiment is also used as a test of theory, or as a device for convincing doubters. He may even declare: 'If you perform such an experiment, then you will obtain such a result,' although he has never made that experiment himself. [... Though experiments] are more common in his books than real ones, for (Galileo thought) we know enough to grasp the truth, if only we learn to reason rightly" (A.R. Hall, *The Scientific Revolution 1500-1800: The Formation of the Modern Scientific Attitude* [Boston: Beacon Press, 1962], 174).

7. Krzysztof Pomian, "Filosofia-Filosofie," in *Enciclopedia Einaudi*, vol. 6 (Turin: Einaudi, 1979).

8. *Ibid.*

9. Alexandre Koyré, *Études d'histoire de la pensée philosophique* (Paris: Colin, 1961), 329.

10. Karl Popper, *Objective Knowledge* (Oxford: Clarendon, 1972), 191.

11. "One cannot deny, as it used to be fashionable to do, the inventiveness and observational capacities of an epoch which built Romanesque basilicas and great Gothic cathedrals, and either invented or adopted and introduced into our material culture objects such as the harness, the horseshoe, stirrups, watermills and windmills, pulleys, compasses, and gunpowder, an epoch during which lenses, mechanical clocks, and scales appeared for the first time." Lucien Febvre, *Le problème de l'incroyance au XV*e *siècle: La religion de Rabelais* (Paris: Albin Michel, 1946), 412. It is also impossible to deny the Middle Ages an interest in the empirical world.

12. Alexandre Koyré, *Études Galiléennes* (Paris: Hermann, 1939), 86-161.

13. Krzysztof Pomian, *Przeszłość jako przedmiot wiary* (Warsaw: 1968).

14. Ernest Gellner, "An Ethics of Cognition," in R.S. Cohen, P.K. Feyerabend and M.W. Wartofsky (eds.), *Essays in Memory of Imre Lakatos*. Boston Studies in the Philosophy of Science, 39 (Dordrecht: Reidel, 1976).

15. Alexandre Koyré, *Études d'histoire de la pensée philosophique*, 320.

16. *Ibid.*, 327.

17. Joseph Agassi, *Towards an Historiography of Science* (Middletown: Wesleyan University Press, 1963).

18. Cited after Paolo Rossi, *Philosophy, Technology and the Arts in the Early Modern Era*, trans. S. Attanasio, ed. B. Nelson (New York: Harper and Row: 1970), 114; emphasis mine.

19. Koyré, *Études d'histoire de la pensée philosophique*, 329.

20. Pierre Francastel, *Peinture et société* (Paris: Denoël/Gauthier, 1951), 206-7.

21. Leszek Kołakowski, *Main Currents of Marxism*, vol. 1, *The Founders* (Oxford: Oxford University Press, 1978).
22. Cited after P. Rossi, *Philosophy, Technology, and the Arts*, 92.
23. Stanisław Ossowski, *Nauka o nauce*, vol. 4, *Dzieła* (Warsaw, PWN, 1967), 100.

Chapter IV. The Institutionalization and Professionalization of Scientific Research

1. Lewis Mumford, *Technics and Civilization* (London: Routledge and Sons Ltd., 1934), 215.
2. Norbert Wiener, *The Human Use of Human Beings: Cybernetics and Society* (New York: Avon Books, 1967), 157.
3. M. Dumas, "Esquisse d'une histoire de la vie scientifique," in *Encyclopédie de la Pléiade: Histoire de la Science* (Paris: Gallimard, 1957), 57.
4. The following numbers testify to the development of the German universities (from J. Ben-David, *The Scientist's Role in Society*, Princeton: Princeton University Press, 1971): The number of students increased from 16,000 in 1876 to 47,000 in 1908; in technical universities, which gained academic status only in 1899, there were 4000 in 1891 and 10,500 in 1899. Let us note that the development of technical universities created an additional demand both for the results of theoretical science and for scientists. They were also becoming the links connecting science with the economy. The size of the faculty increased at a slower rate: from 1313 in 1869 to 2275 in 1890 and over 3000 in 1910. The total university budget in Prussia, Saxony, Bavaria and Württemburg grew from 12 million in 1880 to 40 million in 1914.
5. S. Newcomb, "Exact Sciences in America," in J.C. Burnham (ed.), *Science in America: Historical Selections* (New York: 1971), 204-205.
6. G. Barraclough, *An Introduction to Contemporary History* (Harmondsworth: Penguin Books, 1964), 46-47.
7. Rudolf Hilferding, *Finance Capital: A Study of the Latest Phase of Capitalist Development*, ed. Tom Bottomore, trans. Morris Watnick and Sam Gordon, (London: Routledge & Kegan Paul, 1981 [orig. 1910]), 123.

Chapter V. The Sources of the Crisis of the Modern Ideal of Science

1. E. Ashby, *Technology and the Academics* (London: Macmillan, 1958), 94.
2. See J. Ravetz, *Scientific Knowledge and its Social Problems* (Oxford: Oxford University Press, 1971), 44.
3. The famous British mathematician, G.H. Hardy, wrote: "If useful knowledge is, as we agreed provisionally to say, knowledge which is likely now or in the comparatively near future to contribute to the material comfort of mankind, so that mere intellectual satisfaction is irrelevant, then the great bulk of higher mathematics is useless. [...] If this be the test, then Abel, Riemann and Poincaré wasted their lives. [...] I have never done anything 'useful.' [...] Judged by all practical standards, the value of my mathematics is nil. [...] I have just one chance of escaping a verdict of triviality, that I may be judged to have created something

worth creating" (G.H. Hardy, *A Mathematician's Apology*, Cambridge: Cambridge University Press, 1967, 135-36 and 150).

4. C.P. Snow, *Science and Government* (Cambridge: Harvard University Press, 1961).

5. Descartes, *Discourse on the Method*, in *Philosophical Works of Descartes*, vol. 1, trans. E. Haldane and G.R.T. Ross (Cambridge: Cambridge University Press, 1972), 130.

6. Gerald Holton, "From the Endless Frontier to the Ideology of Limits," in his *The Advancement of Science and Its Burdens* (Cambridge: Cambridge University Press, 1986), 209.

7. Alexandre Koyré, *Newtonian Studies* (Chicago: University of Chicago Press, 1965), 12.

8. Voltaire, *Lettres philosophiques*, in *Mélanges* (Paris: Gallimard, 1960), 112.

9. S. Amsterdamski, "Previsione e possibilità," *Enciclopedia Einaudi*, vol. 10 (Torino: Einaudi, 1980), 1108-1130.

10. Helena Eilstein, "Demon Laplace'owski i gatunek ludzki," *Studia Filozoficzne*, 4 (1966), 125.

11. See S. Amsterdamski, "Naturale/Artificiale," *Enciclopedia Einaudi*, vol. 9 (Torino: Einaudi, 1980), 792-822.

12. Cited after K. Pomian, "Działanie i sumienie," *Studia Filozoficzne*, 1967 (3), 22-23.

13. *Ibid.*, 22-23.

14. *Ibid.*, 22-23.

15. Leszek Kołakowski *The Alienation of Reason: A History of Positivist Thought* (New York: Doubleday, 1968).

16. Barbara Skarga, *Claude Bernard* (Warsaw, 1970), 104.

17. F. A. Hayek, *The Counter-Revolution of Science* (Glencoe, Illinois: Free Press, 1952); H. Schoeck (ed.), *Scientism and Values* (Princeton: Princeton University Press, 1960).

18. Popper, *Objective Knowledge*, 185.

19. A. Roy, *La Philosophie contemporaine* (Paris: 1898), 5.

20. K. Pomian, "Działanie i sumienie," 25.

21. So, while certain authors in Poland in 1968, for example, J. Szewczyk and S. Dziamski, condemned the "scientistic deviation from Marxism" of people who defended these values, other authors elsewhere, such as, for example, A. Hobbs, characterized the scientistic position in the McCarthy Era as having "only a remote relationship with science [...]. Fertilized in the minds of intellectuals and incubated in professional journals, the germs of scientism infect textbooks, and acquire great virulence as they are transmitted to the public [...]. The creed of scientism leads to the belief that our economic system is characterized by maldistribution of income, unemployment, and class conflict. [...] Our adherence to traditional beliefs about child training, education, sex, marriage, economics, religion, patriotism and other matters constitutes culture lag and disorganizes society [...]. The techniques, language, and the creed of scientism, however, are quite similar to those of socialism and communism" (A. Hobbs, *Social Problems and Scientism*, Harrisburg, PA: The Stackpole Co., 1953, 17-19, 37 and *passim*). If one disregards the

positions under attack it is difficult not to notice the similarity of the values underlying this type of critique.

Chapter VI. Escape to World Three

1. Alexandre Koyré, *Études Galiléennes* (Paris: Hermann, 1940), 166-171.
2. Jan Such, *Czy istnieje experimentum crucis?* (Warsaw, 1975); S. Amsterdamski, "Zgoda i niezgoda. Janowi Suchowi w odpowiedzi," *Człowiek i Światopogląd*, 1974, 4.
3. Imre Lakatos, "Falsification and the Methodology of Scientific Research Programmes," in *The Methodology of Scientific Research Programmes. Philosophical Papers*, Vol. 1, (Cambridge: Cambridge University Press, 1978).
4. K. Pomian, "La rationalité, l'irrationalité et la science," *Annales: Économie, Societés, Civilisations*, vol. 30 (5).
5. Karl R. Popper, *Objective Knowledge* (Oxford: Clarendon Press, 1972), and his *Unended Quest* (La Salle, Illinois: Open Court, 1976).
6. Popper, *Unended Quest*, 89.
7. Popper, *Unended Quest*, 148-151.
8. This idea, which can be found in the works of Popper himself, was developed in detail first by Joseph Agassi (in *Towards a Historiography of Science*, The Hague: Mouton, 1963 and his *Science in Flux*, Boston Studies in the Philosophy of Science, Dordrecht: Reidel, 1975) and later, in the context of the newer polemics by I. Lakatos. Contrasting Popper's falsificationism and logical empiricism, Lakatos claimed that as an empirical programme, the latter ceased to be fertile long ago, while never even trying to be a historiographic programme. ("History of Science and its Rational Reconstructions," *The Methodology of Scientific Research Programmes. Philosophical Papers*, Vol. 1, Cambridge: Cambridge University Press, 1978).
9. Karl Popper, *The Logic of Scientific Discovery* (New York: Harper & Row, 1959, 1968), 31.
10. Karl Popper, *Objective Knowledge: An Evolutionary Approach* (Oxford: Clarendon, 1972), 113-114.
11. Popper, *The Logic of Scientific Discovery*, 31-32.
12. Lakatos, "History of Science and its Rational Reconstructions," 102.
13. Popper does not acknowledge this shift, since he interprets *The Logic of Scientific Discovery* according to his current views—not an unusual phenomenon among scholars who have worked in a particular field for over half a century.
14. Karl Popper and John C. Eccles, *The Self and its Brain* (Berlin: Springer International, 1977), 38-39.
15. Popper, *Objective Knowledge*, 159n.
16. *Ibid*, 161.
17. Popper, *Unended Quest*, 183.
18. Popper, *Objective Knowledge*, 80-81.
19. *Ibid.*, 126.
20. *Ibid.*, 261.
21. Popper, *Unended Quest*, 169.

22. Popper, *Objective Knowledge*, 70.
23. Popper and Eccles, *The Self and its Brain*, 48.
24. Karl Popper, *Conjectures and Refutations* (London: Routledge and Kegan Paul, 1963), 352.
25. *Ibid.*, 361.

Chapter VII. Are There Selection Criteria?

1. On the distinction between formal and semantic correspondence see my *Between Experience and Metaphysics* (Dordrecht: Reidel, 1975), ch. 7.
2. *Ibid.*, ch. 4.
3. I ignore here the issue of whether the degree of confirmation can be expressed quantitatively, as Carnap believed possible (see *Logical Foundations of Probability*, Chicago: University of Chicago Press, 1950). It is well known that a higher degree of justification cannot be equated with a higher degree of truth (as Reichenbach wanted); and because of this, a selection criterion which requires that we choose the theory which appears the most probable, given the available information, is not based on the degree of truth, nor can it be considered conclusive.
4. Pierre Duhem, *The Aim and Structure of Physical Theory* (Princeton: Princeton University Press, 1954), 187.
5. Both of the above theses are frequently attributed to Duhem, and sometimes they are even conflated with each other. Since what is at stake is not a correct interpretation of Duhem's position but the analysis of his thesis, I find it necessary, like many other authors writing on the subject, to distinguish clearly between these two theses.
6. Adolf Grünbaum, "Falsifiability and Rationality," paper read at the *International Colloquium on Issues in Contemporary Physics and Philosophy of Science*, September 1971, manuscript.
7. It is perhaps worth noting that despite some interpretations of the experiment of Pasteur, it is not only the rejection of the theory of spontaneous generation, but also its justification, which could constitute an argument in controversies about the supernatural origin of life.
8. Jan Such, *Czy istnieje experimentum crucis?* (Warsaw: 1975), 531.
9. *Ibid.*, 527, see also p. 50 and note.
10. Falsification is regarded as a process also by Lakatos, in "Falsification and the Methodology of Scientific Research Programmes"; however, in *The Methodology of Scientific Research Programmes. Philosophical Papers*, Vol. 1 (Cambridge: Cambridge University Press, 1978), he does not identify the problem of the crucial experiment with the entirety of this process, but only with the issue of whether there is a last element such that after its occurrence, the theory is definitively falsified.
11. Such, *Czy istnieje experimentum crucis?*, 98-99.
12. *Ibid.*, 99.
13. Rejection of radical empiricism excludes the possibility of testing isolated hypotheses, whereas its acceptance allows for both possibilities.

14. Karl Popper, *Conjectures and Refutations* (London: Routledge and Kegan Paul, 1963), 112; Lakatos, "Falsification and the Methodology of Scientific Research Programmes."

15. Karl Popper, *The Logic of Scientific Discovery* (New York: Harper and Row, 1959), 78n.

16. This is the position defended by Adolf Grünbaum in "The Falsifiability of a Component of a Theoretical System," in *Mind, Matter and Method*, eds. P.K. Feyerabend and G. Maxwell (Minneapolis: University of Minnesota Press, 1966). In his later work Grünbaum abandoned this view and basically accepted Duhem's thesis. See his "Can We Ascertain the Falsity of a Scientific Hypothesis?" in *Studium Generale*, 22 (1969), 1061-93; or "Falsifiability and Rationality," and "*Ad hoc* Auxiliary Hypotheses and Falsificationism," *British Journal for the Philosophy of Science*, 27 (1976). Concerning Grünbaum's position, see also Lakatos, "Falsification and the Methodology of Scientific Research Programs," 187, and part 6 of this chapter.

17. Accordingly Duhem's thesis is sometimes understood as if it excluded the possibility of falsifying an isolated hypothesis, and sometimes as if it excluded falsification of any isolated fragment of knowledge. I believe that whoever accepts the thesis in this first formulation must also accept the second.

18. Willard Van Orman Quine, *From a Logical Point of View: Logico-Philosophical Essays* (New York: Harper & Row, 1953, 2nd rev. ed. 1961), 43.

19. *Ibid.*, 44.

20. Grünbaum, "The Falsifiability of a Component of a Theoretical System," 280.

21. *Ibid.*, and his "Can We Ascertain the Falsity of A Scientific Hypothesis?"

22. Grünbaum, "The Fasificability of a Component of a Theoretical System," 278.

23. Grünbaum, "*Ad hoc* Auxiliary Hypotheses and Falsificationism," §3.

24. Quine, *From the Logical Point of View*, 44.

25. Lakatos, "Falsification and the Methodology of Scientific Research Programmes," 93.

26. Popper, *Conjectures and Refutations*, 38n.

27. Stefan Amsterdamski, *Between Experience and Metaphysics* (Dordrecht: Reidel, 1975), ch. 6.

28. Grünbaum, "*Ad hoc* Auxiliary Hypotheses and Falsificationism," 350.

29. Carl Hempel, *Philosophy of Science* (New York: Prentice-Hall, 1966), 30.

30. Grünbaum, "*Ad hoc* Auxiliary Hypothesis and Falsificationism"; Popper, *Unended Quest*.

31. Joseph Agassi, *Science in Flux*, Boston Studies in the Philosophy of Science (Dordrecht: Reidel, 1975).

32. The following statement by John Watkins in defense of Popperian philosophy nicely illustrates how difficult it sometimes is to decide where continuation ends and essential revision begins: "Yes, we have no criteria. What we do have is a corrigible and revisable methodology of scientific appraisal" (John Watkins, "The Popperian Approach to Scientific Knowledge," in G. Radnitzky and G. Andersson [eds.], *Progress and Rationality in Science* [Dordrecht: Reidel, 1978], 3). This is supposed to be a *continuation* of Popper's position, using his criteria of demarcation and falsification.

33. Lakatos, "Falsification and the Methodology of Scientific Research Programmes."
34. My views on this subject have been presented in *Between Experience and Metaphysics*, chapter 6. Today I would add to this evaluation (given the arguments presented in the previous section) that I doubt whether criteria of selection can be unequivocal even within the framework of a single programme. (See also Alan Musgrave, "Method or Madness?" in R.S. Cohen, P.K. Feyerabend and M.W. Wartofsky (eds.), *Essays in Memory of Imre Lakatos* (Dordrecht: Reidel, 1976).
35. In a paper written together with Elie Zachar shortly before his death, Lakatos defended this position quite explicitly ("Why did Copernicus's Programme Supersede Ptolemy's?" in Imre Lakatos, *The Methodology of Scientific Research Programmes* (Cambridge: Cambridge University Press, 1978).
36. Musgrave, "Method or Madness?"
37. Lakatos, "History of Science and Its Rational Reconstructions," in *The Methodology of Scientific Research Programmes*, 112, 117.
38. *Ibid.*, 103n.
39. Lakatos, "Falsification and the Methodology of Scientific Research Programmes," ch. II.
40. Ph. Quinn, "Metaphysical Appraisal and Heuristic Advice: Problems in Methodology of Scientific Research Programmes," in *Studies in History and Philosophy of Science*, vol. 3 (1972), 143.
41. Lakatos, "Changes in the Problem of Inductive Logic," in *Mathematics, Science, and Epistemology. Philosophical Papers*, Vol. 2, eds. John Worrall and Gregory Currie (Cambridge: Cambridge University Press, 1978), 147.
42. Imre Lakatos, "Reply to Critics," in R.C. Buck and R.S. Cohen (eds.), *P.S.A. 1970, Boston Studies in the Philosophy of Science*, 8 (Dordrecht: Reidel, 1971), 174-82.
43. Grünbaum, "Falsifiability and Rationality," 89-90.
44. Since quasi-falsification is not conclusive, Grünbaum faces the same problem as Lakatos and Musgrave: how to reconcile the heuristic significance of a methodology with the fact that it does not supply unequivocal selection criteria.
45. Musgrave, "Method or Madness," 476, 479-480.
46. Lakatos, "History of Science and Its Rational Reconstructions," 117.
47. *Ibid.*, 117, 133.

Chapter VIII. Order and Anarchy

1. T.S. Kuhn, *The Essential Tension* (Chicago: University of Chicago Press, 1977), ch. 13.
2. T.S. Kuhn, "Postscript—1969" to *The Structure of Scientific Revolutions* (2nd ed., Chicago: University of Chicago Press, 1970).
3. Paul K. Feyerabend, *Against Method: Outline of an Anarchistic Theory of Knowledge* (London: NLB, 1975), 10.
4. Paul K. Feyerabend, "The Methodology of Scientific Research Programmes," in *Problems of Empiricism. Philosophical Papers*, Vol. 2 (Cambridge: Cambridge University Press, 1981), 203. See also "On the Critique of Scientific Reason," in R.S. Cohen, P.K. Feyerabend, and M.W. Wartofsky (eds.), *Essays in Memory of Imre Lakatos* (Dordrecht: Reidel, 1976) and his "In Defense of Aristotle:

Comments on the Condition of Content Increase," in Gerard Radnitzky and Gunnar Andersson (eds.), *Progress and Rationality in Science* (Dordrecht: Reidel, 1978).

5. Krystyna Zamiara, "Wstęp" [Introduction] to P.K. Feyerabend, *Jak być dobrym empirystą* (Warsaw: PWN, 1979) 5.

6. Paul K. Feyerabend, "How to Be a Good Empiricist: A Plea for Tolerance in Matters Epistemological," in *Philosophy of Science: The Delaware Seminar*, Vol. 2 (New York: Wiley Interscience, 1963), 8.

7. As far as I know, neither Feyerabend nor Kuhn, who reached an identical position on this issue at approximately the same time, were aware of the work of Kazimierz Ajdukiewicz from the 1930s, in which Ajdukiewicz presented his view of closed and internally consistent languages; see Kazimierz Ajdukiewicz, "Das Weltbild und die Begriffsapparatur," *Erkenntnis*, vol. 4 (1934), 259-287; and his "Sprache und Sinn," *Erkenntnis*, vol. 4 (1934), 100-138. They also never analyzed the reasons which led Ajdukiewicz to abandon up this position with respect to the language of science. Both authors cite Benjamin Lee Whorf and Edward Sapir as their sources of inspiration.

8. Feyerabend, "How to Be a Good Empiricist," 10.

9. *Ibid.*, 6.

10. *Ibid.*, 6.

11. *Ibid.*, 37.

12. "Wherever facts play a role in such a dogmatic defense, we shall have to suspect foul play [...] the foul play of those who try to turn good science into bad, because unchangeable, metaphysics." *Ibid.*, 38.

13. Still, in the summary of the eighteenth chapter of *Against Method*, we can read: "Thus science is much closer to myth than a scientific philosophy is prepared to admit. It is one of the many forms of thought that have been developed by man, and not necessarily the best. [... It is the] most recent, most aggressive and most dogmatic religious institution" (Feyerabend, *Against Method*, 295).

14. Feyerabend, "How to be a good empiricist," 7.

15. *Ibid.*, 6-7.

16. *Ibid.*, 30.

17. Feyerabend, "Problems of Empiricism," in *Beyond the Edge of Certainty*, ed. R.G. Colodny (Englewood Cliffs: Prentice-Hall, 1965), ch. 1.

18. Popper, "Normal Science and its Dangers," in Imre Lakatos and Alan Musgrave (eds.), *Criticism and the Growth of Knowledge* (Cambridge: Cambridge University Press, 1970).

19. Paul K. Feyerabend, *Rationalism and Scientific Method. Philosophical Papers*, Vol. 1 (Cambridge: Cambridge University Press, 1981), 94.

20. *Ibid.*, 44ff.

21. Paul K. Feyerabend, "Consolations for a Specialist" in *Problems of Empiricism. Philosophical Papers*, Vol. 2 (Cambridge: Cambridge University Press, 1981) also his "In Defense of Aristotle."

22. Those authors who argue against radical empiricism on the one hand, and on the other defend the principle of correspondence in the sense discussed above, simply have not drawn all the theoretical consequences which follow from the rejection of radical empiricism. See for example W. Krajewski, *The Correspondence*

Principle and the Growth of Science (Dordrecht: Reidel, 1977); and W. Krajewski, W. Mejbaum, J. Such (eds.), *Zasada korespondencji w fizyce a rozwój nauki*, (Warsaw: PWN, 1974).

23. "Today the world is messages, codes and information. Tomorrow what analysis will break down our objects to reconstitute the in a new space? What new Russian doll will emerge?" François Jacob, *The Logic of Life: A History of Heredity*, tr. Betty E. Spillmann (New York: Random House, 1973), 324. Each such regrouping is a change of the semantic model of our vision of the world or of its fragment.

24. Feyerabend, "Problems of Empiricism," 169.

25. *Ibid.*, 169.

26. An analogy between the language of scientific theories with Ajdukiewicz's model of a closed and internally consistent language, such that an introduction of each new term changes the meaning of all known terms, was regarded as a fiction by Ajdukiewicz himself. Such languages can be constructed but they do not correspond or "closely approximate" theories of the empirical sciences.

27. See Adolf Grünbaum, "Can a Theory Answer More Questions Than One of Its Rivals?" *British Journal for the Philosophy of Science*, 27 (1976).

28. Feyerabend, *Against Method*, 298.

29. Feyerabend, "In Defense of Aristotle," 173.

30. Feyerabend, *Against Method*, chapter 16; his "On the Critique of Scientific Reason," in R.S. Cohen. P.K. Feyerabend and M. Wartofsky (eds.), *Essays in Memory of Imre Lakatos* (Dordrecht: Reidel, 1978) 110; and "In Defense of Aristotle," 153 and *passim*.

31. John Watkins, "Corroboration and the Problem of Content Comparison," in G. Radnitzky and G. Andersson (eds.), *Progress and Rationality in Science* (Dordrecht: Reidel, 1978), 339.

32. Feyerabend, "In Defense of Aristotle," 174.

33. *Ibid.*, 173.

34. Feyerabend, *Against Method*, 187.

Appendix: Philosophy of Science and Sociology of Knowledge

1. The present text is the original version of a paper I presented at the Wissenschaftskolleg zu Berlin on June 14, 1988. Invited to publish it in the *Jahrbuch*, I added a short Postscript taking into account some of the comments made during the two discussions in the Kolleg.

2. J. R. Brown (ed.), *Scientific Rationality: The Sociological Turn* (Dordrecht: Reidel, 1981).

3. I use the term introduced by Leszek Kołakowski in his essay "The Epistemological Significance of the Aetiology of Knowledge: A Gloss on Mannheim," *TriQuarterly*, Fall 1971.

4. M. Norton Wise, "Mediating Machines," *Science in Context*, 2 (1988), 77-113; M. N. Wise and C. Smith, "Measurement, Work and Industry in Lord Kelvin's England," *Historical Studies in the Physical and Biological Sciences*, 17 (1986), 147-73.

5. John Farley and Gerald Geison, "Science, Politics and Spontaneous Generation in Nineteenth Century France: The Pasteur-Pouchet Debate," *Bulletin of the History of Medicine*, 48 (1974), 161-198.
6. Paul Forman, "Weimar Culture, Causality and Quantum Theory 1918-1927," *Historical Studies in the Physical Sciences*, 3 (1971), 1-114.
7. See the letter to *The New York Review of Books*, 22 (1975), 18.
8. T.S. Kuhn, *The Structure of Scientific Revolutions*, (Chicago: University of Chicago Press, 1970), 9.
9. S. Amsterdamski, "Le concept du sujet cognitive et l'évolution de la science," *Fundamenta Scientiæ*, 6 (1985), 313-325.
10. I mean Popper as we know him from *The Logic of Scientific Discovery*, but not from *Objective Knowledge*.
11. Popper, *Objective Knowledge*, especially chapter entitled "Epistemology without the Knowing Subject."
12. Alexandre Koyré, "De l'influence des concéptions philosophiques sur l'évolution des théories scientifiques" in *Études d'histoire de la pensée philosophique* (Paris: Colin, 1961), 236.
13. This remark does not apply to Peter Galison, who differentiates long-term constraints in experimental as well as in theoretical investigations. For example he writes: "Such presuppositions offer an analogue to Braudel's geographical time, for they are not attached to the goals of any single research group and frequently not even to a single scientific specialty. [...] Such beliefs lasted for centuries. Sometimes, as Gerald Holton has argued, transcultural commitments may come in "thematic pairs" such as the belief that nature must be explained in terms of continuous or in terms of discrete matter" (*How Experiments End* (Chicago: University of Chicago Press 1987), 247). But even Galison does not notice that the epistemological significance of his long-term constraints is not the same as the significance of middle- and short-term constraints.
14. Bruno Latour and Steven Woolgar, *Laboratory Life: The Social Construction of a Scientific Fact* (Beverly Hills: Sage, 1979) and Bruno Latour, *Science in Action* (Milton Keynes: Open University Press, 1989).
15. Timothy Lenoir, "Introduction" to *Science in Context*, 2 (1988) and Wise, "Mediating Machines."
16. *Ibid.*, 4.
17. *Ibid.*
18. Galison, *How Experiments End*, 277.
19. For the critique and defense of the intellectual history of science see A. Lugg "Two Historical Strategies: Ideas and Social Conditions in the History of Science," in Brown (ed.), *Scientific Rationality: The Sociological Turn*.
20. Ian Hacking, *Representing and Intervening* (Cambridge: Cambridge University Press, 1983), 274.
21. Wise, "Mediating Machines," 109 [italics mine—S.A.].
22. *Ibid.*
23. Richard Whitley *The Intellectual and Social Organization of the Sciences* (Oxford: Clarendon, 1984), 11-12.
24. Hacking, *Representing and Intervening*, 275.

25. Henri Poincaré, *Les sciences et les humanités* (Paris: 1908), 31.
26. Wise, "Mediating Machines," 107-110.
27. *Ibid.*, 108.
28. *Ibid.*, 108.
29. See Alexandre Koyré, "Du monde de l'à-peu-près à l'univers de la précision," *Études d'histoire de la pensée scientifique* (Paris: P.U.F., 1961), 311-329.
30. Lugg, "Two Historiographical Traditions."
31. Both arguments were advanced by Norton Wise in his "Rebuttal" to my text which he read during the discussion; for an edited text see M. Norton Wise, "Practice: A Missing Link in History and Philosophy of Science," *Jahrbuch 1987/1988: Wissenschaftskolleg.* Institute for Advanced Study, Berlin (Berlin: Nicolaische Verlagbuchhandlung, 1989), 174.

BIBLIOGRAPHY

Agassi, Joseph. *Towards a Historiography of Science*. The Hague: Mouton, 1963.

-------. *Science in Flux*. Boston Studies in the Philosophy of Science. Dordrecht: Reidel, 1975.

Ajdukiewicz, Kazimierz. "Das Weltbild und die Begriffsapparatur." *Erkenntnis*, 4 (1934), 259-287.

-------. "Sprache und Sinn." *Erkenntnis*, 4 (1934), 100-134.

Amsterdamski, Stefan. "Posłowie" [Editor's Postscript] to the Polish translation of T.S. Kuhn, *The Structure of Scientific Revolutions*. Trans. H. Ostromęcka. Warsaw: 1968.

-------. "Scjentyzm wczoraj i dziś" [Scientism Yesterday and Today]. In *Z Historii Filozofii Pozytywistycznej w Polsce. Ciągłości i Przemiany* [History of Positivist Thought in Poland. Continuities and Change]. Ed. A. Hochfeld and B. Skarga. Warsaw: Ossolineum, 1972.

-------. "Scjentyzm a rewolucja naukowo-techniczna" [Scientism and the Technological and Scientific Revolution]. *Zagadnienia naukoznawstwa*, 23:3 (1970), 16-33.

-------. "Zgoda i niezgoda. Janowi Suchowi w odpowiedzi" ["Agreement and Disagreement. A Reply to Jan Such"]. *Człowiek i Światopogląd*, 4 (1974).

-------. *Between Experience and Metaphysics*. Dordrecht: Reidel, 1975.

-------. "Lege." *Enciclopedia Einaudi*. Vol. 8. Turin: Einaudi, 1979, 83-129.

-------. "Naturale/artificiale." *Enciclopedia Einaudi*. Vol. 9. Turin: Einaudi, 1980, 792-822.

-------. "Previsione e possibilita." *Enciclopedia Einaudi*. Vol. 10. Turin: Einaudi, 1980, 1108-1130.

-------. "Ripetizione." *Enciclopedia Einaudi*. Vol. 12. Turin: Einaudi, 1981, 76-86.

-------. *Nauka a porządek świata* [Science and Order of the World]. Warsaw: PWN, 1983.

-------. "Le concept du sujet cognitive et l'évolution de la science." *Fundamenta Scientiæ*, 6 (1985), 313-325.

Ashby, E. *Technology and the Academics*. London: Macmillan, 1958.

Barraclough, G. *An Introduction to Contemporary History*. Harmondsworth: Penguin Books, 1964.

Ben-David, Joseph. *The Scientist's Role in Society*. Princeton: Princeton University Press, 1971.

Berdhal, R.O. *British Universities and the State*. Berkeley: University of California Press, 1959.

Bernal, J. D. *The Social Function of Science*. London: Routledge & Kegan Paul, 1939.

Braithwaite, R. *Scientific Explanation*. Cambridge, Cambridge University Press: 1955.

Bridgman, P. W. *The Logic of Modern Physics*. New York: Macmillan, 1927.

Brown, J. R. (ed.). *Scientific Rationality: The Sociological Turn*. Dordrecht: Reidel, 1981.

Butterfield, Herbert. *The Origins of Modern Science, 1300-1800*. London: Bell & Hyman, 1949.

--------. *History and Human Relations*. London: Collins, 1951.

Carnap, Rudolf. *Logical Foundations of Probability*. Chicago: Chicago University Press, 1950.

-------. "Testability and Signs." *Philosophy of Signs*, 3 (1936); 4 (1937).

Chmielewska, E. *Kategorie 'kontekst odkrycia' i 'kontekst uzasadnienia' we współczesnej filozofii nauki* [The Categories 'Context of Discovery' and 'Context of Justification' in Contemporary Philosophy of Science]. Ph.D. Diss. Warsaw University, 1978.

Conant, J. B. *Modern Science and Modern Man*. New York: Columbia University Press, 1952.

Descartes, René. *Discourse on the Method of Rightly Conducting the Reason*. In *The Philosophical Works of Descartes*. Trans. E. Haldane and G.R.T. Ross. Cambridge: Cambridge University Press, 1972.

Donagan, Alan. "The Popper-Hempel Theory Reconsidered." In *Philosophical Analysis and History*. Ed. William H. Dray. Westport, CT: Greenwood Press, 1966.

Duhem, Pierre. *The Aim and Structure of Physical Theory*. Princeton: Princeton University Press, 1954.

Dumas, M. "Esquisse d'une histoire de la vie scientifique." In *Encyclopédie de la Pléiade: Histoire de la Science*. Paris: Gallimard, 1957.

Eilstein, Helena. "Czas-możliwość" [Time-Possibility]. *Studia Filozoficzne*, 1961, no. 4, 3-26.

-------. "Demon Laplace'owski i gatunek ludzki." *Studia Filozoficzne*, 1966, no. 4.

Elkana, Yehuda. "Cultural Systems and Science." In *Essays in Memory of Imre Lakatos*. Eds. R.S. Cohen, P.K. Feyerabend, and M.W. Wartofsky. Boston Studies in Philosophy of Science, 39. Dordrecht: Reidel, 1976.

-------. *The Problem of Knowledge in Historical Perspective*. Athens, 1973.

Ellul, Jacques. *The Technological Society*. New York: Vintage Press, 1964.

Engel, A. "The Emerging Concept of Academic Profession in Oxford, 1980-1854." In *The University in Society*. Ed. Lawrence Stone. Princeton: Princeton University Press, 1974.

Evans-Pritchard, E.E. *Social Anthropology*. London: Cohen & West, 1951.

Farley, John and Gerald Geison. "Science, Politics and Spontaneous Generation in Nineteenth-Century France: The Pasteur-Pouchet Debate." *Bulletin of the History of Medicine*, 48 (1974), 161-198.

Febvre, Lucien. *Le Problème de l'incroyance au XV* siècle: La religion de Rabelais*. Paris: Albin Michel, 1946.

Feyerabend, Paul K. "Explanation, Reduction, and Empiricism." In *Scientific Explanation, Space and Time*. Eds. H. Feigl and G. Maxwell. Minnesota Studies in the Philosophy of Science, 3. Minneapolis: Minnesota University Press, 1962, 28-97. [Also in his *Realism, Rationalism and Scientific Method: Philosophical Papers*. Vol. 1. Cambridge: Cambridge University Press, 1981.]

-------. "How to Be a Good Empiricist: A Plea for Tolerance in Matters Epistemological." In *Philosophy of Science: The Delaware Seminar*. Vol. 2. New York: Wiley Interscience, 1963.

-------. "Problems of Empiricism, Part I. In *Beyond the Edge of Certainty: Essays in Contemporary Science and Philosophy*. Ed. Robert G. Colodny. Englewood Cliffs, N.J.: Prentice-Hall, 1965.

-------. "Problems of Empiricism, Part II." In *The Nature and Function of Scientific Theories*. Ed. Robert G. Colodny. Pittsburgh: University of Pittsburgh Press, 1970.

-------. *Against Method: Outline of an Anarchistic Theory of Knowledge*. New York: NLB, 1975.

-------. "On the Critique of Scientific Reason." In *Essays in Memory of Imre Lakatos*. Eds. R.S. Cohen, P.K. Feyerabend, and M.W. Wartofsky. Boston Studies in Philosophy of Science, 39. Dordrecht: Reidel, 1976.

-------. "In Defence of Aristotle: Comments on the Condition of Content Increase." In *Progress and Rationality in Science*. Eds. Gerard Radnitzky and Gunnar Andersson. Boston Studies in Philosophy of Science, 58. Dordrecht: Reidel, 1978.

--------. *Science in a Free Society*. New York: NLB, 1978.

--------. "The Methodology of Scientific Research Programmes." In his *Problems of Empiricism. Philosophical Papers*. Vol. 2. Cambridge: Cambridge University Press, 1981.

-------. "Consolations for the Specialist," in his *Problems of Empiricism. Philosophical Papers*. Vol. 2. Cambridge: Cambridge University Press, 1981.

Fontenelle, B. "Digression sur les Anciens et les Modernes." In *Œuvres*. Vol. 4. Paris: 1767.

Forman, Paul. "Weimar Culture, Causality and Quantum Theory 1918-1927: Adaptation by German Physicists and Mathematicians to a Hostile Intellectual Environment," *Historical Studies in the Physical Sciences*, 3 (1971), 1-114.

Francastel, Pierre. *Peinture et societé: naissance et destruction d'un espace plastique*. Paris: Denoël/Gauthier, 1977.

Galileo Galilei. *Two Chief World Systems*. Trans. S. Drake. Berkeley: University of California Press, 1967.

Galison, Peter. *How Experiments End*. Chicago: University of Chicago Press, 1987.

Gellner, Ernest. *Legitimation of Belief*. Cambridge: Cambridge University Press, 1974.

-------. "An Ethics of Cognition." In *Essays in the Memory of Imre Lakatos*. Eds. R.S. Cohen, P.K. Feyerabend, and M.W. Wartofsky. Boston Studies in the Philosophy of Science, 39. Dordrecht: Reidel, 1976.

Giedymin, Jerzy. "Indukcjonizm i antyindukcjonizm" [Inductivism and anti-inductivism]. In *Logiczna teoria nauki*. Ed. T. Pawłowski. Warsaw: PWN, 1966.

-------. "Revolutionary Changes, Non-Translatability and Crucial Experiments." In *Problems in the Philosophy of Science*. Eds. I. Lakatos and A. Musgrave. International Colloquium in the Philosophy of Science, Bedford College, 1965, Proceedings. Vol. 3. Amsterdam: North-Holland Publishing Co., 1968.

-------. "Instrumentalism and Its Critique: A Reappraisal." In *Essays in Memory of Imre Lakatos*. Eds. R.S. Cohen, P.K. Feyerabend, and M.W. Wartofsky. Boston Studies in the Philosophy of Science, 39. Dordrecht: Reidel, 1976.

Grünbaum, Adolf. "Law and Convention in Physical Theory." In *Current Issues in the Philosophy of Science*. Eds. Herbert Feigl and G. Maxwell. New York: Holt, Rinehart and Winston, 1961.

-------. "The Falsifiability of a Component of a Theoretical System." In *Mind, Matter and Method*. Eds. P.K. Feyerabend, G. Maxwell. Minneapolis: University of Minnesota Press, 1966.

-------. "Can We Ascertain the Falsity of a Scientific Hypothesis?" In *Studium Generale*, 22 (1969), 1061-93.

-------. "Falsifiability and Rationality." Paper read at the *International Colloquium on Issues in Contemporary Physics and Philosophy of Science*, Pennsylvania State University, September 1971, manuscript.

-------. "*Ad hoc* Auxiliary Hypotheses and Falsificationism." *The British Journal for the Philosophy of Science*, 27 (1976).

-------. "Can a Theory Answer More Questions Than One of Its Rivals?" *The British Journal for the Philosophy of Science*, 27 (1976), 1-23.

-------. "Is Falsifiability the Touchstone of Scientific Rationality?" In *Essays in Memory of Imre Lakatos*. Eds. R.S. Cohen, P.K. Feyerabend, and M.W. Wartofsky. Boston Studies in the Philosophy of Science, 39. Dordrecht: Reidel, 1976.

-------. "Popper *versus* Inductivism." In *Progress and Rationality in Science*. Eds. G. Radnitzky and G. Andersson. Boston Studies in Philosophy of Science, 58. Dordrecht: Reidel, 1978.

Habermas, Jürgen. "Technology and Science as 'Ideology.'" In his *Toward a Rational Society: Student Protest, Science, and Politics*. Trans. Jeremy J. Shapiro. London: Heinemann, 1971.

-------. *Knowledge and Human Interests*. Trans. Jeremy J. Shapiro. London: Heinemann, 1972.

Hacking, Ian. *Representing and Intervening*. Cambridge: Cambridge University Press, 1983.

Hall, A. R. *The Scientific Revolution 1500-1800: The Formation of the Modern Scientific Attitude*. Boston: Beacon Press, 1962.

Hardy, G.H. *A Mathematician's Apology*. Cambridge: Cambridge University Press, 1967.

Hayek, F.A. *The Counter-Revolution of Science*. Glencoe, Illinois: Free Press, 1952.

-------. *New Studies in Philosophy, Politics, Economics and the History of Ideas*. London: Routledge and Kegan Paul, 1978.

Hazard, Paul. *The European Mind: 1680-1715*. New Haven: Yale University Press, 1953.

-------. *La Pensée européenne au XVIIIe siècle, de Montesquieu à Lessing*. Paris: Fayard, 1978 [orig. 1946].

Hempel, Carl G. "The Function of General Laws in History." *Journal of Philosophy*, 39 (1942).

-------. "Deductive-Nomological versus Statistical Explanations." In *Scientific Explanation, Space and Time*. Eds. H. Feigl and G. Maxwell. Minnesota Studies in the Philosophy of Science, 3. Minneapolis: Minnesota University Press, 1962.

-------. "Explanation in Science and History." In *Frontiers of Science and Philosophy*. Ed. R. Colodny. Pittsburgh: University of Pittsburgh Press, 1962.

-------. *Philosophy of Science*. New York: Prentice Hall, 1966.

------- and P. Oppenheim. "Studies in the Logic of Explanation." *Philosophy of Science*, 15 (1948).

Hilferding, Rudolf. *Finance Capital: A Study of the Latest Phase of Capitalist Development*. Ed. Tom Bottomore. Trans. Morris Watnick and Sam Gordon. London: Routledge & Kegan Paul, 1981 [orig. 1910].

Hobbs, A. *Social Problems and Scientism*. Harrisburg, PA: The Stackpole Co., 1953.

Holton, Gerald, ed. *Science and Culture: A Study of Cohesive and Disjunctive Forces*. Boston: Beacon Press, 1965.

-------. *Thematic Origins of Scientific Thought: Kepler to Einstein.* Cambridge: Harvard University Press, 1973.

-------. "From the Endless Frontier to the Ideology of Limits." In his *The Advancement of Science and Its Burdens.* Cambridge: Cambridge University Press, 1986.

Jacob, François. *The Logic of Life: A History of Heredity.* Tr. Betty E. Spillmann. New York: Random House, 1973.

Kmita, Jerzy. "Słowo wstępne" [Introduction] to the Polish translation of K. R. Popper *The Logic of Scientific Discovery.* Trans. U. Niklas. Warsaw, 1977.

Kołakowski, Leszek. *Kultura i fetysze.* Warsaw: PWN, 1967.

-------. *The Alienation of Reason: A History of Positivist Thought.* New York: Doubleday, 1968.

-------."The Epistemological Significance of the Aetiology of Knowledge: A Gloss on Mannheim." *TriQuarterly*, Fall 1971.

-------. *Obecność mitu.* Paris: Instytut Literacki, 1976.

-------. *Husserl and the Search for Certainty.* New Haven: Yale University Press, 1975.

-------. *Main Currents of Marxism: Its Rise, Growth and Dissolution.* Trans. P.S. Falla. Oxford: Clarendon Press, 1978.

Koyré, Alexandre. *Études Galiléennes.* Paris: Hermann, 1939.

-------. *From the Closed World to the Infinite Universe.* Baltimore: The Johns Hopkins University Press, 1957.

-------. *Études d'histoire de la pensée philosophique.* Paris: Colin, 1961.

-------. *Newtonian Studies.* Chicago: Chicago University Press, 1965.

-------. *Études d'histoire de la pensée scientifique.* Paris: P.U.F., 1966.

Krajewski, W. *The Correspondence Principle and the Growth of Science.* Dordrecht: Reidel, 1977.

-------, W. Mejbaum, and J. Such (eds.). *Zasada korespondencji w fizyce a rozwój nauki* [The Correspondence Principle in Physics and the Development of Science]. Warsaw: PWN, 1974.

Kuhn, Thomas S. *The Copernican Revolution: Planetary Astronomy in the Devel-opment of Western Thought.* Cambridge: Harvard University Press, 1957.

-------. *The Structure of Scientific Revolutions.* Chicago: University of Chicago Press, 1962.

-------. "Postscript" to the 2nd edition of *The Structure of Scientific Revolutions.* Chicago: University of Chicago Press, 1970.

--------. *The Essential Tension.* Chicago: University of Chicago Press, 1977.

Lakatos, Imre. "Proofs and Refutations." *The British Journal for the Philosophy of Science,* 14 (1963-64).

-------. "History of Science and Its Rational Reconstructions." In *The Methodology of Scientific Research Programmes. Philosophical Papers.* Vol. 1. Eds. John Worrall and Gregory Currie. Cambridge: Cambridge University Press, 1978.

-------. "Falsification and the Methodology of Scientific Research Programmes." In *The Methodology of Scientific Research Programmes. Philosophical Papers.* Vol. 1. Eds. J. Worrall and G. Currie. Cambridge: Cambridge University Press, 1978.

-------. "Changes in the Problem of Inductive Logic." In *Mathematics, Science, and Epistemology. Philosophical Papers.* Vol. 2. Eds. J. Worrall and G. Currie. Cambridge: Cambridge University Press, 1978.

-------. "Reply to Critics." In *P.S.A. 1970.* Eds. R.C. Buck and R.S. Cohen. Boston Studies in the Philosophy of Science, 8. Dordrecht: Reidel, 1971.

------- and Elie Zachar. "Why did Copernicus's Programme Supersede Ptolemy's?" In *The Methodology of Scientific Research Programmes. Philosophical Papers.* Vol. 1. Eds. J. Worrall and G. Currie. Cambridge: Cambridge University Press, 1978.

Latour, Bruno. *Science in Action: How to Follow Scientists and Engineers Through Society.* Milton Keynes: Open University Press, 1989.

------- and Steven Woolgar. *Laboratory Life: The Social Construction of a Scientific Fact.* Beverly Hills: Sage, 1979.

Laudan, Larry. *Progress and its Problems.* Berkeley: University of California Press, 1977.

Lenoir, Timothy. "Introduction" to *Science in Context,* 2 (1988).

Leplin, Jarret. "The Concept of an *ad hoc* Hypothesis." *Studies in the History and Philosophy of Science*, 5, 1975.

Lévi-Strauss, Claude. *Structural Anthropology.* Trans. Claire Jacobson, B.G. Schoef and Monique Layton. Harmondsworth: Penguin Books, 1977/78.

--------. "Réponses à quelques questions." *Esprit*, 31 (11 Nov. 1963).

A. Lugg. "Two Historiographical Traditions." In *Scientific Rationality: The Sociological Turn*. Ed. J.R. Brown. Dordrecht: Reidel, 1981.

Magee, Bryan. *Karl Popper.* New York: Viking Press, 1973.

Marcuse, Herbert. *One-Dimensional Man.* Boston: Beacon Press, 1964.

Mendelski, T. *Karl Popper: metodolog czy ideolog?* [Karl R. Popper: Methodologist or Ideologue?]. Warsaw: 1978.

Merton, Robert K. *Science, Technology and Society in 17th Century England.* New York: Harper and Row, 1970 [orig. 1938].

-------. *Social Theory and Social Structure.* New York: The Free Press, 1957.

Mokrzycki, Edmund. *Philosophy of Science and Sociology.* London: Routledge and Kegan Paul, 1983.

Mumford, Lewis. *Technics and Civilization.* London: Routledge and Sons Ltd., 1934.

Musgrave, Alan. "Method or Madness?" In *Essays in Memory of Imre Lakatos*. Eds. R.S. Cohen, P.K. Feyerabend, and M.W. Wartofsky. Boston Studies in Philosophy of Science, 39. Dordrecht: Reidel, 1976.

-------. "Evidential Support, Falsification, Heuristics and Anarchism." In *Progress and Rationality in Science*. Eds. G. Radnitzky and G. Andersson. Boston Studies in Philosophy of Science, 58. Dordrecht: Reidel, 1978.

Nagel, Ernest. *The Structure of Science.* New York: Harcourt, Brace and World, 1961.

Newcomb, S. "Exact Sciences in America." In *Science in America: Historical Selections*. Ed. J.C. Burnham. New York: 1971.

Ortega y Gasset, J. "Man the Technician." In *History as a System*. New York: 1962.

Ossowski, Stanisław. *O Nauce* [On Science]. *Dzieła*. Vol. 4. Warsaw: PWN, 1967.

Panofsky, E. "Galileo as a Critic of the Arts: Aesthetic Attitude and Scientific Thought." *Isis*, 47 (1956), 3-15.

Pietruska-Madej, E. *W poszukiwaniu praw rozwoju nauki* [In Search of Laws of Scientific Development]. Warsaw: PWN, 1980.

Poincaré, Henri. *Les sciences et les humanités*. Paris: Flammarion, 1908.

-------. *The Foundations of Science*. Trans. G.B. Halsted. Lancaster, PA: The Science Press, 1913.

Pomian, Krzysztof. "Działanie i Sumienie" [Action and Conscience]. *Studia Filozo ficzne*, 1967 (3).

--------. *Przeszłość jako przedmiot wiary* [The Past as an Object of Faith]. Warsaw: PWN, 1968.

-------. "Rozum w świecie paradoksu" [Reason in the World of Paradox]. In his *Człowiek pośród rzeczy* [Man Among Things]. Warsaw: Czytelnik, 1974.

-------. "Kartezjusz: Negatywność indywiduum i nieskończoność nauki" [Descartes: Negativity of an Individual and the Infinity of Science]. In his *Człowiek pośród rzeczy* [Man Among Things]. Warsaw: Czytelnik, 1974.

--------. "Filosofia-Filosofie." In *Enciclopedia Einaudi*. Vol. 6. Torino: Einaudi, 1979.

--------. "La rationalité, irrationalité et la science." *Annales: Économie, Société, Civilisations*, 30 (5).

-------. "Les différentes conceptions de la connaissance de la nature." Paris: C.N.R.S., 1980 (manuscript).

Popper, Karl R. *The Logic of Scientific Discovery*. New York: Harper and Row, 1959.

-------. *Conjectures and Refutations*. London: Routledge and Kegan Paul, 1963.

--------. "Normal Science and Its Dangers." In *Criticism and the Growth of Knowledge*. Eds. I. Lakatos and A. Musgrave. Cambridge: Cambridge University Press, 1970.

--------. *Objective Knowledge: An Evolutionary Approach*. Oxford: Clarendon Press, 1972.

-------. *Unended Quest: An Intellectual Autobiography*. La Salle, Ill.: Open Court, 1976.

------- and John C. Eccles. *The Self and its Brain*. Berlin: Springer International, 1977.

Price, D.K. *Government and Science.* New York: Oxford University press, 1962.

-------. *The Scientific Estate.* Cambridge: Harvard University Press, 1965.

Price, Derek de Solla. *Little Science, Big Science.* New York: Columbia University Press, 1962.

-------. *Science Since Babylon.* New Haven: Yale University Press, 1975.

Quine, Willard Van Orman. *From a Logical Point of View: Logico-Philosophical Essays.* New York: Harper & Row, 1953, 2nd rev. ed. 1961.

Quinn, Ph. "Metaphysical Appraisal and Heuristic Advice: Problems in Methodology of Scientific Research Programmes." In *Studies in History and Philosophy of Science.*" Vol. 3 (1972).

Radnitzky, Gerard. "The Popperian Philosophy of Science as an Antidote Against Relativism." In *Essays in Memory of Imre Lakatos.* Eds. R.S. Cohen, P.K. Feyerabend, and M.W. Wartofsky. Boston Studies in Philosophy of Science, 39. Dordrecht: Reidel, 1976.

-------. From the Logic of Science to the Theory of Research." *Communications,* 7 (1974).

------- and Gunnar Andersson. "Objective Criteria of Scienitfic Progress?" In *Progress and Rationality in Science.* Eds. G. Radnitzky and G. Andersson. Boston Studies in Philosophy of Science, 58. Dordrecht: Reidel, 1978.

Ravetz, J. *Scientific Knowledge and its Social Problems.* Oxford: Oxford University Press, 1971.

Ricoeur, Paul. "Structure et herméneutique." *Esprit* (November 1963).

Rose, Hilary and Steven Rose. "The Incorporation of Science." 1974 (mimeographed text).

Roszak, Theodore. *Where the Wasteland Ends.* Garden City, NY: Anchor Books, 1972.

Rossi, Paolo. *Philosophy, Technology and the Arts in the Early Modern Era.* Trans. S. Attanasio. Ed. B. Nelson. New York: Harper and Row, 1970

Roy, A. *La Philosophie contemporaine.* Paris: 1898.

Schoeck, H. S. (ed.). *Scientism and Values.* Princeton: Princeton University Press, 1960.

Schumacher, E.F. *Small is Beautiful: A Study of Economics as if People Mattered.* London: Blond & Briggs, 1973.

Scriven, Michael. "Explanations, Predictions and Laws." In *Scientific Explanation, Space and Time.* Eds. H. Feigl and G. Maxwell. Minnesota Studies in the Philosophy of Science, 3. Minneapolis: Minnesota University Press, 1962.

-------. "Causes, Connections and Conditions in History." In *Philosophical Analysis and History.* Ed. William H. Dray. Westport, CT: Greenwood Press, 1966.

Skarga, Barbara. *Claude Bernard.* Warsaw: Wiedza Powszechna, 1970.

-------. "Porządek świata i porządek wiedzy" [The Order of the World and the Order of Knowledge]. In *Z Historii Filozofii Pozytywistycznej w Polsce: Ciągłości i Przemiany* [A History of Positivist Thought in Poland: Continuities and Changes]. Ed. A. Hochfeld and B. Skarga. Warsaw: Ossolineum, 1972.

Snow, C. P. *Science and Government.* Cambridge: Harvard University Press, 1961.

Storer, Norman. *The Social System of Science.* New York: Holt, Rinehart and Winston, 1966.

Such, Jan. *Czy istnieje experimentum crucis? Problemy sprawdzania praw i teorii naukowych* [Do Crucial Experiments Exist? Problems of Testing Scientific Laws and Theories]. Warsaw: PWN, 1975.

-------. "Racje i zastrzeżenia" [Reasons and Objections]. *Człowiek i Światopogląd,* (1974, 4).

Thom, René. *Stabilité structurelle et morphogenèse: Essai d'une théorie générale des modèles.* Paris: InterEditions, 1977.

Toulmin, Stephen. *The Uses of Argument.* Cambridge: Cambridge University Press, 1958.

-------. *Human Understanding: The Collective Use and Evolution of Concepts.* Princeton, Princeton University Press, 1972.

Tyrowicz, S. *Światło wiedzy zdeprawowanej.* Poznań: 1970.

Unger, P. *Ignorance: A Case for Scepticism.* Oxford: Oxford University Press, 1975.

Urbach, P. "The Objective Promise of a Research Program." In *Progress and Rationality in Science.* Eds. G. Radnitzky and G. Andersson. Boston Studies in Philosophy of Science, 58. Dordrecht: Reidel, 1978.

Voltaire. *Lettres philosophiques. Mélanges.* Paris: Gallimard, 1960.

Watkins, John. "The Popperian Approach to Scientific Knowledge." In *Progress and Rationality in Science.* Eds. G. Radnitzky and G. Andersson. Dordrecht: Reidel, 1978.

-------. "Corroboration and the Problem of Content Comparison." In *Progress and Rationality in Science.* Eds. G. Radnitzky and G. Andersson. Dordrecht: Reidel, 1978.

-------. "Against Normal Science." In *Criticism and the Growth of Knowledge.* Eds. I. Lakatos and A. Musgrave. Cambridge: Cambridge University Press, 1970.

Weizsäcker, Carl Friedrich. *Die Einheit der Natur.* München: Carl Hansen, 1971.

Whitley, Richard. *The Intellectual and Social Organization of the Sciences.* Oxford: Clarendon, 1984.

Wiener, Norbert. *The Human Use of Human Beings: Cybernetics and Society.* New York: Avon Books, 1967.

Wise, M. Norton. "Mediating Machines." *Science in Context,* 2 (1988), 77-113.

-------. "Practice: A Missing Link in History and Philosophy of Science," *Jahrbuch 1987/1988. Wissenschaftskolleg.* Institute for Advanced Study, Berlin. Berlin: Nicolaische Verlagbuchhandlung, 1989.

------- and C. Smith, "Measurement, Work and Industry in Lord Kelvin's England." *Historical Studies in the Physical and Biological Sciences,* 17 (1986), 147-73.

Worrall, J. "The Ways in which the Methodology of Scientific Research Programmes Improves on Popper's Methodology." In *Progress and Rationality in Science.* Eds. G. Radnitzky and G. Andersson. Dordrecht: Reidel, 1978.

-------. "Research programmes, Empirical Support and the Duhem Problem." In *Progress and Rationality in Science.* Eds. G. Radnitzky and G. Andersson. Dordrecht: Reidel, 1978.

Zachar, E. "Crucial Experiments." In *Progress and Rationality in Science.* Eds. G. Radnitzky and G. Andersson. Dordrecht: Reidel, 1978.

Zamiara, Krystyna. "Wstęp" [Introduction] to P.K. Feyerabend, *Jak być dobrym empirystą* [How to Be A Good Empiricist]. Warsaw: PWN, 1979.

Zilsel, Edgar. "Physics and the Problem of Historico-sociological Laws." *Philosophy of Science,* 8 (1941), 567-580.

Ziman, John. *Public Knowledge: The Social Dimension of Science*. London: Cambridge University Press, 1968.

Znaniecki, Florian. *The Social Role of the Man of Knowledge*. New York: Columbia University Press, 1940.

INDEX OF NAMES